RSPB
Bird Tales

Traditional Stories, Folklore and Activities

Dawn Casey
Illustrated by Lucinda Warner

BLOOMSBURY WILDLIFE
LONDON · OXFORD · NEW YORK · NEW DELHI · SYDNEY

With deep gratitude, to Lucinda
DAWN CASEY

For Rowan and Robin, my own little birds
LUCINDA WARNER

BLOOMSBURY WILDLIFE
Bloomsbury Publishing Plc
50 Bedford Square, London, WC1B 3DP, UK
29 Earlsfort Terrace, Dublin 2, Ireland

BLOOMSBURY, BLOOMSBURY WILDLIFE and the
Diana logo are trademarks of Bloomsbury Publishing Plc

This edition published in the United Kingdom 2024
Text © Dawn Casey 2024
Illustrations © Lucinda Warner 2024

The extract from April Birthday by Ted Hughes
on page 94, and the extract from Swifts on page 105,
are both © Ted Hughes. They are reproduced by
kind permission of Faber and Faber Ltd.

ISBN: HB: 978-1-3994-0690-1
ePub: 978-1-3994-0691-8
ePDF: 978-1-3994-0692-5

2 4 6 8 10 9 7 5 3 1

Designed by Austin Taylor
Printed and bound in China by RR Donnelley
Asia Printing Solutions Ltd, Dongguan, Guangdong

FSC MIX Paper | Supporting responsible forestry FSC® C144853

To find out more about our authors and books visit
www.bloomsbury.com and sign up for our newsletters.

RSPB

Published under licence from RSPB Sales Limited to
raise awareness of the RSPB (charity registration in England
and Wales no 207076 and Scotland no SC037654).

For all licensed products sold by Bloomsbury Publishing
Limited, Bloomsbury Publishing Limited will donate a
minimum of 2% from all sales to RSPB Sales Ltd, which gives
all of its distributable profits through Gift Aid to the RSPB.

CONTENTS

PREFACE

The egg is a universal symbol of the miracle of life. The nest is a marvel. Birds can *fly*! Yet these wonders are part of everyday life – the ducks in the park, the Kestrel above the road, the Robin in the garden. Birds grace our days.

At a challenging time in my life, the Blackbird singing outside my bedroom window reminded me every day that life is beautiful. A few years later, a female Blackbird trained me to bring her mealworms. Every time I came outside, she flew over to remind me. I feel the fondest love for the birds. And I love the old stories.

In modern cultures, information has often been valued over imagination. Yet in indigenous cultures across the world, *stories*, rather than facts alone, helped people learn, and remember, real truths. Stories work with the imagination and the feelings – the ways of relating to the world that come naturally to children. A child might look at a picture of a Great Spotted Woodpecker in a reference book and soon forget the bird's markings. But hearing the folktale of the old woman with the white apron and the red cap gives children a picture that stays in their memory. Relating to a bird in a story can be a first step to creating a sense of kinship and developing a real friendship.

When my children were little, I looked for books of bird stories to share. The collections I found were full of tales of Peacocks and Phoenixes – beautiful, magical birds – but I wanted to give them stories of the birds we knew from our garden. These are the tales I have sought out to share with you here. Some are familiar favourites, many are little known. All are crafted for clarity – they are easy to remember and easy to tell aloud.

There are stories, crafts and activities for children to enjoy with parents and grandparents, and by themselves. There are rhymes and songs to share with even the littlest ones.

I also include a wealth of folklore – the wisdom and wonder of our ancestors, crystalised into bright gems. In selecting what to include, I have chosen folklore that celebrates the birds, and our own connections with them. I have chosen folklore that relates directly to how each bird looks and lives – old lore that can help us get to know and love each bird, here and now.

Bird Tales is a resource for children and for all those who guide children in connecting with nature – families, teachers, storytellers and forest school leaders. I hope it will help and inspire both the young and the young at heart, to deepen their relationships with the birds.

Even in the Ice Age, our ancestors carved the rocks with pictures of birds. In the British Isles, we have a long and rich tradition of appreciating our feathered friends. With humble hand and loving heart I add this offering to those that came before.

Dawn Casey

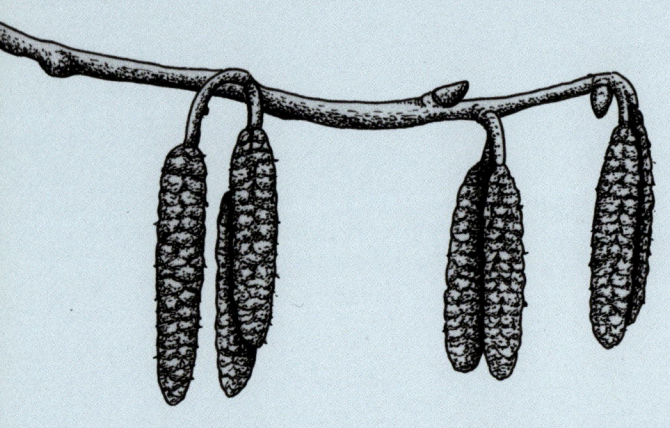

January – the blank page at the start of the new year.
The bare trees make silhouette shapes upon the rose glow of the sky, like intricate, exquisite papercuts. Often there is frost, framing the edges of the fallen leaves, the curves of the pines, the terracotta tiles on the rooftops. The ice earth scrunches beneath our boots. Muddy puddles are frozen solid. The cold air on our faces makes our skin sing. Hazel catkins shiver in the wind.

Winter is a good time to begin getting to know the birds, when flocks are a spectacular sight, patterning the winter skies, and birds are easy to see amidst the bare branches. Members of the crow family (corvids) are especially easy to spot. They are large and loud, and live alongside humans, even in towns and cities.

The corvid family includes the familiar Magpie, the colourful Jay, the huge Raven and the red-billed Chough.

In this chapter, we meet three of the most commonly seen black crows, the Carrion Crow, the Rook and the Jackdaw.

JANUARY

JACKDAW

Coloeus monedula

Grey 'hood'

Pale eyes

Juveniles have duller
plumage and a darker nape.

VOICE
The Jackdaw says its name:
tchack-tchack. Jackdaw flight-calls
can be very loud – the collective
noun for a group of Jackdaws is
a clattering. Its 'crow's caw'
is a squeaky, clipped *kya*.

SIZE
34cm

WHERE
Open woodland,
farmland, towns.

NEST
Hole-nest, often
in colony, same
site used year
after year. Male
brings female
food whilst she
sits on eggs.

EATS
Almost anything.

JACKDAW – THE OLD NAMES

Names echoing the Jackdaw's cry:
Jack • Jakie (Scotland) **• Jac-y-do** (Wales) **• Kya** (Wigtownshire)
Ka (Orkney Isles) **• Daw • Cawdaw** (Suffolk) **• Caddaw** (East Anglia)

From the Jackdaw's grey nape:
Grey Head/Grey Neck (Midlothian) **• Grey Pate** (Kirkcudbright)

LITTLE JACK

Like Robin Redbreast and Jenny Wren, Jack is an affectionate name for a familiar friend. The name Jack has often been given to little ones (Jack Snipe, for example, is smaller than Common Snipe, and the jack is the smallest-value picture card in a pack of cards). The Jackdaw is the smallest of the black corvids, distinctly smaller than the Rook or the Carrion Crow. The jack in a pack of playing cards is also known as the knave, and the Jackdaw, too, is associated with a knave's roguish ways. Its scientific name, *Coloeus monedula*, stems from the Greek *moneta* ('money') – they were thought to collect shiny objects, like their cousins the 'thieving Magpies'. There's a grain of truth in the myth. Jackdaws' friendly nature and remarkable intelligence meant that they were sometimes reared as pets, and even taught to imitate human speech. Crows of all kinds were such popular pets in early Ireland that there was a law stating that they must be kept on a string, to stop them getting into mischief. In Italy, thieves once famously trained tame Jackdaws to steal money from people at cash machines. Young Jackdaws peck at any small object that contrasts with its background, as they learn what is, and isn't, good to eat. In the countryside, young Jackdaws play with all kinds of found treasures, like sticks, stones and acorn cups, while in urban areas, a Jackdaw's playthings might include human-made finds, such as a shiny ring pull. Perhaps humans are simply more likely to notice a Jackdaw picking up something when it is shiny – the old belief may tell us more about our own ways of seeing than about the bird's behaviour.

CHURCHES AND CHIMNEY POTS

Jackdaws nest in cavities, in holes in old trees, cliffs or in the ruins of old buildings. In towns and cities, chimney pots offer a suitable alternative. To make a nest, a pair drops sticks into their chosen hole until a few lodge, creating a platform to build upon.

To make a warm nest lining, a Jackdaw will sometimes perch on the back of a cow, horse or sheep (or even a wild deer) to pluck out tufts of fur. They also eat the bugs found on the animal's back (as well as those found in dried-up cowpats), so look out for Jackdaws in fields of cattle or sheep.

When every home had a fireplace, Jackdaws were sometimes considered a nuisance, as a Jackdaw's nest could block a chimney and even crash down into the fireplace, with a Jackdaw still in it. When a Jackdaw fell into my (luckily unlit) hearth, it looked me right in the eye, and waited patiently for me to open the window before flying to freedom.

Jackdaws also often make their nests in church towers, a connection which meant the bird was held sacred in some parts of Wales.

There is a bird who, by his coat
And by the hoarseness of his note,
Might be supposed a crow;
A great frequenter of the church,
Where, bishop-like, he finds a perch,
And dormitory too.

EXTRACT FROM 'THE JACKDAW'
BY WILLIAM COWPER

DEVOTED PARTNERS

Listen for the male Jackdaw loudly announcing his arrival at the nest, and the female's soft reply. Young Jackdaws pair up during the spring, and stay together for a whole year before starting a family. Like other crows, Jackdaws mate for life. While most birds (around 90 per cent) are monogamous, having only one mate within a breeding season, Jackdaws stay with the same partner not just for one breeding season, but through the winter too, year after year, staying together even if their early breeding attempts do not result in a brood.

Even when Jackdaws join a flock, partners stay together – notice pairs sitting side by side amidst the crowd. They strengthen their bond by preening the feathers on their partner's head and neck. The birds' relationship informed an old belief that Jackdaws are an omen of a happy marriage.

Jackdaws' contact calls have a comfortable, companionable sound, as they chat softly together on the chimney pots.

WISE HEART

The corvid family are renowned for their intelligence. Jackdaws can use tools and can recognise individual human faces. So far, only a few animals are known to be able to do this (including other corvids, chimpanzees and pigeons). Studies by zoologist Auguste von Bayern showed that Jackdaws respond to human facial expressions. Like humans, Jackdaws use eye contact to communicate.

A Jackdaw can 'read' a human's eyes, following a mentor's eye movements to find hidden food. Tamed Jackdaws have been taught to say human words, parrot-fashion.

Jackdaws will care for an injured relative and actively share food (as well as other objects) with other Jackdaws, even those outside their own families. They also share more of their 'favourite' foods than other foods. Jackdaws' fabled cleverness also includes a high level of emotional intelligence.

'THE JACKDAW' A TALE INSPIRED BY AESOP'S FABLE

A story celebrating the distinctive features of the Jackdaw

Once upon a chimney pot, a Jackdaw sat. From high above the rooftops, he could see a patchwork of fields and farms, deep lanes and high hedgerows, and cottages all in a row. In the grounds of the big house – the one with lots of chimney pots – Peacocks strutted, trailing their long bright feathers like grand evening gowns. How glamorous they looked. How glorious!

The little Jackdaw had never paid too much attention to how his feathers looked, but now he looked down at his plain black coat and he felt dull.

With swift wings the Jackdaw swooped down to the ground. Long luxurious Peacock plumes lay upon the grass, fallen feathers that the birds had cast off. Long luscious curves, tipped with a dazzling eye, gleaming emerald green, lapis blue, antique gold.

The Jackdaw pecked at a feather with his beak. He picked it up and stuck it amongst his own feathers. He found another. And another. Soon the Jackdaw wore Peacock plumes sticking out in all directions.

But the proud Peacocks, when they saw the Jackdaw, squawked 'Those are *our* feathers!', and they pecked him and plucked the plumes away with sharp beaks. They stalked off across the lawn, their long necks held high, without a backward glance at the bedraggled Jackdaw.

The Jackdaw plodded away. He came to the pond and looked into the water. He saw his own ruffled feathers, his plain colours, his dull coat. He sighed.

But then, the wind blew. It fluttered the Jackdaw's feathers, it whispered to him, it tugged at him. The Jackdaw spread his wings and answered the call. The cool air rippled through his feathers, carrying away his cares. He beat his wings with swift strokes, savouring his speed. A tilt, and he could arc and swing, riding the curves of the wind. And when he held his wings wide, he could glide through the air, sliding on the sky.

At last he landed back by the pond, breathless, bright-eyed. He looked at his reflection in the water and he saw himself afresh: he saw his striking light eyes, his smart grey hood. And his beautiful black wings – his wonderful, wind-soaring wings.

ROOK

Corvus frugilegus

Bare patch of skin at base of beak –
makes beak look long and pale

Glossy black plumage – in
some lights, body looks purple
or silver-blue; wings and tail
may have a green sheen

Thickly feathered
thighs, like
'baggy trousers'

Juveniles have darker bills and no bare
face patch, but can be distinguished from
Carrion Crows by their 'baggy trousers'.

VOICE
A Rook's long drawn-out
kraa-ah has a softer, more
mellow tone than a crow's deep,
forceful *caw*. Often several Rook
voices are heard at once.

SIZE
45cm

WHERE
Pasture,
farmland,
woodland,
villages.

NEST
Rooks nest
communally, in
a rookery – a
group of bulky
stick nests, high
up in tall trees.

EATS
Almost
anything, but
particularly
loves acorns.

ROOK – THE OLD NAMES

The Rook's scientific name stems from the Latin *frugilegus* meaning 'fruit-picking'.
Both Rooks and Carrion Crows were often called, simply, 'Crow':
Cra (Westmorland) • Craw (North Yorkshire, Lancashire)

Names echoing the Rook's cry:
Hroc (Anglo-Saxon) • **Croaker**

From the Rook's distinctive face:
Bare-faced Crow (Scotland) • White-faced Crow (Scotland)

A CROWD OF CROWS

A Crow in a crowd is a Rook,
A Rook on its own is a Crow.
COUNTRY SAYING

Most of the time, the old saying above holds true. Rooks are the most sociable of the corvid family, nesting together, feeding together and roosting together. In winter, they flock together in groups that can contain hundreds (sometimes even thousands) of birds.

On winter evenings, Rooks return to a communal roost, using the same site each night, often deep in a wood. Birds from several rookeries travel for miles (using set sky-paths) to gather there (and are often joined by Jackdaws, and sometimes Carrion Crows too). The shared winter roost could be one colony's rookery, or other tall trees nearby.

Knowledge of a favourite roost is passed down from one generation to the next. Sometimes when trees without nests are used, the site was once a rookery – though the nests have gone, the connection to place lives on.

'January'
Whilst many a mingled swarthy crowd, —
Rook, crow, and jackdaw, — noising loud,
Fly to and fro to dreary fen,
Dull Winter's weary flight again;
They flop on heavy wings away
As soon as morning wakens grey,
And, when the sun sets round and red,
Return to naked woods to bed.
FROM *THE SHEPHERD'S CALENDAR*, JOHN CLARE

There's something eternal about Rooks returning to the roost, like the cycle of the year or the rhythm of the day – a sense that the Earth continues to turn and all is well with the world.

The number of Rooks roosting together reaches its peak in January. If you live near woods or tall trees, take a moment to venture into your garden on a January afternoon before sunset. There's a good chance you'll be treated to the sight of Rooks journeying towards bed.

A ragged ribbon of Rooks, clear black shapes upon the deep pink sunset, is a quintessential sight on a winter's afternoon.

THE ROOKERY

Within the rookery, young Rooks are more protected within a large group of adults. If one bird spots danger and sounds the alarm, all the others drive off the intruder together.

On the feast day of beautiful Bride
The flocks are counted on the moor.
The raven goes to prepare the nest,
And again goes the rook,
Nest at Brigit, egg at Shrove, chick at Easter.

ORAL LORE COLLECTED FROM ISLE OF UIST
BY ALEXANDER CARMICHAEL

The 'feast day of beautiful Bride', also known as the feast day of St Brigit, is the 1st February – Rooks are one of the earliest species to breed. From February onwards, look for Rooks carrying (often very large) sticks in their beaks, a sign that nest maintenance has begun. Pairs of Rooks use the same rookery each year, repairing an old nest by building on top of it or using sticks from any surviving nests from the year before to make a new nest. Sometimes, a bird might help itself to a stick from a neighbour's nest-in-process – listen to the birds' noisy caws as they sort out any differences of opinion.

In March, the rookery is alive with noisy activity. The branches are still bare – with a keen eye, or a pair of binoculars, it's easy to watch the Rooks' fascinating world, as the birds talk and flirt, nest and feed, play and settle disagreements together. In March, Rooks roost in their nest trees too. Dawn or dusk – those magical moments that frame the day – are especially good times to visit.

In early spring, watch the male court his mate with a gentlemanly bow, and a display of wing drooping and tail fanning. A pair of Rooks feeds each other, preens each other and even mirrors each other, one bird copying its mate's movements.

As soon as they've made their nest, the female lays her eggs and sits to keep them warm. Both in courtship and whilst the female is incubating, the male brings food to his mate. Watch how she receives his offerings with tremulous voice and fluttering wings.

Because they breed so early in the year, the eggs may be chipping open by the end of March – perfectly timed for the warm wet weather that brings an abundance of earthworms, the staple diet of young Rooks. The adults carry the food to their young in their throat pouches. Later in the year, when the earth is drier, worms burrow deeper down. Long periods of dry summer weather mean Rooks can't get to the food in the hard ground, hence the saying, 'as hungry as a June Crow'.

Once youngsters have made their own way in the world (around June), the rookery is visited less often, though in winter Rooks may meet there before flying together to roost.

Some rookeries contain just a handful of nests; others as many as 30 or 40. The ancient rookery in Crow Wood at Hatton Castle, Aberdeenshire, once held around 9,000 nests. It is still Britain's largest rookery and the largest land bird colony in Britain.

Rookeries are ancestral lands, passed down through the centuries. A rookery, like a historic building, an ancient castle or a stately home, is home to the same family through the generations, forming a living link between the distant past and the present day. This passing on of traditional knowledge from one generation to the next could even be called culture.

As home-loving humans we can feel a deep sense of relatedness to the way Rooks live. In the country, often every village has a rookery. No wonder that the presence of a rookery near your home is good luck. In places as far apart as Cornwall and Northumberland, it was said that if Rooks nesting on your land left the rookery, someone in the family (or even the whole family) would soon also leave the land. The fortunes of human folk and bird folk are indeed entwined.

ROOK

15

PLAYING

Rooks are playful, as well as sociable. They play with sticks and stones and you may even see two Rooks playing tug-of-war. On warm autumn days, if you hear a chorus of Rooks cawing, look for a flock spiralling upwards on outstretched wings, riding a thermal of warm air. They rise with ease, and at the top of the curl of wind, they dive, head-long, zigzagging towards the ground. Because the birds often dive down in pairs, these flights were known in rural communities as 'crows' weddings'. On blustery autumn days, their noisy acrobatics are an exhilarating sight. W. B. Yeats's poem, 'The Cold Heaven', speaks of 'the rook-delighting Heaven'. To human eyes, it does indeed look as if Rooks are delighting in the wind.

TRAILING THE TRACTOR

Look for Rooks on farmland, where they search out worms, grubs and root crops.

In autumn, after the crops are gathered in, the tractor ploughs the yellow stubble into rows of rich brown earth. Ploughing can be seen in autumn and winter months (and even into spring). Traditionally, 'plough Monday' is the first Monday after Epiphany (6th January). It was the day when the farming year began again, after the twelve days of Christmas, with the start of the winter ploughing season.

A tractor ploughing a field often trails a flock of Rooks (as well as gulls and Jackdaws) in its wake. As the plough turns over the earth, it also turns up grubs like the leatherjacket – the larva of the daddy longlegs – providing Rooks with a favourite feast. Since leatherjackets eat root vegetables and the roots of plants, they can be harmful to crops, so the farmer and the Rook each benefit from the bird's tractor-trailing ways.

Rooks are ground-feeders; they plunge their long pointed beaks deep into the soil in search of worms, insects and grubs. The bare patch at the base of their beaks helps them dig deep and stops their feathers getting matted with mud.

ROOK WEATHER LORE

When Rooks nest high, expect a fine, calm summer. But if they build low, rain and wind will follow. Many Rook behaviours were said to foretell rain:

- 'Coming home' to the rookery or roosting site early (Devon and Lancashire).
- Feeding in the streets of the village (Durham).
- Flying to the hills (Isle of Man).
- A tumbling flight or dropping in flight (Devon).

RELIGIOUS ROOKS

Rooks often nest in churchyards. Look for rook-eries in tall trees around country churches. Perhaps

this explains why there are so many old beliefs about Rooks as religious birds. The nursery rhyme 'Who killed Cock Robin?', for example, gives the Rook the role of parson:

> *Who'll be the parson?*
> *I, said the Rook,*
> *With my little book,*
> *I'll be the parson.*

According to one legend, corvids were once white birds but turned black whilst mourning the crucifixion. Another belief held that if you did not wear something new on Easter day, the Rooks would 'spoil your clothes'. In Shropshire, it was thought that Rooks did not work on Sundays, or on Ascension Day, sitting quietly in the trees instead.

AS THE ROOK FLIES

The expression 'as the crow flies', meaning the shortest, most direct route, most likely refers to the Rook, which flies in a direct line, especially when returning to its roost.

SCARECROW

> *One for the rook,*
> *One for the crow,*
> *One to rot,*
> *One to grow.*

SUFFOLK FARMER'S RHYME

As the old rhyme shows, farmers expected corvids to eat half their crop! Scarecrows were made from turnips (or, like Worzel Gummidge, from mangel-wurzels) and old clothes stuffed with straw. The flocks of 'crows' they were made to scare away were probably Rooks, as it's Rooks that are seen in great crowds feeding on farm fields (though they may well be joined by a few crows). In past centuries, village boys and the farmer's young children were employed at seed-times, and again when grain was ripening, as bird-scarers ('crow-keepers', 'crow-herds' or 'bird-boys'). It was a child's first job, much as delivering newspapers might be today. Children were sent to the fields with wooden rattles (sometimes simply pebbles in a tin), clappers or flags to keep the birds off the seed. They used whatever was to hand to scare off the birds, hence the expression 'stone the crows'. Right up to the end of the nineteenth century, crow-herds sang special songs to shoo the birds away:

> *We've ploughed our land,*
> *We've sown our seed,*
> *We've made all neat and gay.*
> *So take a bit, and leave a bit,*
> *Away, birds, away!*

WINTER'S END

From their faithful flight lines and tall treetop roosts, Rooks in winter catch a glimpse of the subtle signs of the turning of the year, as yet unseen by human eyes. Their nest 'speculating', sometimes begun when only the snowdrops dare brave the cold, is a heartening sight.

> *Over the land half freckled with snow*
> *half-thawed*
> *The speculating rooks at their nests cawed,*
> *And saw from elm-tops, delicate as a flower*
> *of grass,*
> *What we below could not see, Winter pass.*

'THAW', EDWARD THOMAS

'THE ROOKS AND THE PEAR TREES' A BRITISH FOLKTALE

A story celebrating the way Rooks live in community

In the days when waggons and drays went rumbling along the old road, there lived a carter. He liked to be out on the road, with the clip-clop rhythm of the horse's hooves and the dear wild cries of the birds. He liked the keen cool rush of the air on his face, the warmth of the rain on his skin and the good smell of the earth when he breathed in. He went trundling through meadow and village, his cart loaded with leather or linen, flour or malt.

After he'd been going up that old road a time or two, he noticed five fine pear trees sat in the hedge that grew alongside the road. In the springtime, the pear trees were a froth of blossom, white clouds against the broad blue sky. In summer, the leaves were glossy green. And come September, thereabouts, they were loaded with ripe pears, glowing tawny gold in the afternoon sun. The carter whoa-ed up his horses and picked himself a pocketful of pears. They were lovely. Soft as butter, sweet as treacle.

Well, there was an inn along the road, where tired travellers and weary horses could rest awhile. One day, the carter was sitting in the inn, and as he sat there, sipping his drink, he got to wondering how five fine pear trees got planted in a hedgerow, right alongside the high road. The carter asked the innkeeper. The innkeeper smiled and nodded his old head, and this is the tale that he told:

Well, those pear trees first grew in a rich meadow, up by Pershore. There was some old elm trees near 'em, with a rookery high up in the branches. The Rooks would come and settle in the branches of the pear trees and peck-peck-peck they'd go. The man that owned the land was fed up of the Rooks eating *his* pears. But truth be told, there's been a rookery on that land since before he was born, and the Rooks had always shared the pears, with the Jackdaws and the Starlings, and the human folk.

Well one day, it were warm and sunny, and the fruit mellow ripe. So *kyaw!* goes one old Rook, then *kyaw!* goes another, and then *flip-flop*, the whole hustle of 'em settles in the pear tree. The man sees the Rooks perching in the pear trees, talking to one another, *krr-up … krr-up …* And he was that fed up, he says to himself, 'I'll get my old gun and I'll shoot the lot of 'em!' So off he goes and back he comes with his gun loaded. *Bang! Bang-bang!*

Kraaa! The Rooks rose up into the air, all together. But, they didn't let go of the trees. No, they closed their toes around the branches and set up a-cawing and a-flapping fit to bust. They made such a *tremenjous* flapping that up come all five pear trees by the roots, and *whiffley-whiffley-whiffley* off they flies, pear trees and all, over along the river towards Tewkesbury. And all that old man could do was stand and stare.

After they'd been a-going a while, *clunk, clunk, clunk,* they drops down the pear trees into a hedge. Took good root they did too, in that fine sandy soil, and there they be, to this very day.

Well, those pear trees flourished, and so do all the folk – human folk and feathered folk – that share their pears.

START A BIRD JOURNAL

January is the perfect time to start a nature diary. Keeping a bird journal for a whole year builds a picture of birdlife season by season, all year round. It can become a precious personal record of a particular year.

Carrying a notebook to record birds you see helps you remember details, so that later you can identify species with a reference guide. Back home, you can write up your notes in your bird journal. Taking notes in the field and keeping a bird journal both fix details into our long-term memory, and sharpen the senses. Sketching a bird makes us look at it more closely – the more we look, the more we see. To record a bird they've seen in a bird journal, children can stick in their notebook sketches or copy (or trace) a picture from a field guide to colour. All three processes require close observation and attention to detail, helping us learn and remember specific features of the bird.

A bird journal can include:

- **Date.**
  Time of year is key – some birds, like Swallows, come and go with the seasons.
- **Place.**
  Where was the bird seen or heard? Some birds are only found in certain habitats. You could draw or stick a map into your book.
- **Weather.**
- **The appearance of the bird.**
  Size, colour, markings.
- **What the bird is doing.**
  How does it move and feed? Is it out in the open or hiding in undergrowth? Alone or with others? To estimate how many birds are in a flock, count 10 birds, then estimate how many groups of 10 are in the flock.
- **Call/song.**
- **A sketch of the bird.**
- **Anything else you want to record and remember.**

Rook feather

Pair of Jackdaws

28th JANUARY 4.30pm
Kingston Village Church
– Cold + Still
As the light faded watched flocks of rooks and jackdaws flying in to roost. About 10 rooks and 20 jackdaws. Noisy 'tchack tchacking.'
Saw pairs of jackdaws sitting side by side. ♡

CARRION CROW AND HOODED CROW

Corvus corone and *Corvus cornix*

CARRION CROW

All-black feathers

Heavy black beak

Juveniles of both are browner than adults.

HOODED CROW

Black 'hood'

Grey back

Grey breast

VOICE
A crow's low croaking caw is often repeated three times: *kraaa … kraaa … kraaa …* The calls of Carrion Crows and Hooded Crows are almost indistinguishable.

SIZE
45cm

WHERE
Open country, parks, gardens.

NEST
Bulky stick nest. Male brings female food while she incubates eggs.

EATS
Almost anything.

CARRION CROW AND HOODED CROW – THE OLD NAMES

Names echoing the Carrion Crow's throaty, vibrant call:
Crāwe (Anglo-Saxon) • **Craw** (Scotland) • **Corbie Crow** (Northern England)

The Carrion Crow eats carrion, hence:
Flesh Crow (Yorkshire) • **Ket Crow** – 'ket' means carrion (Northern England)

The Carrion Crow scavenges rubbish tips, hence:
Midden Crow – 'midden' means muckheap • **Gor Crow** – from Anglo-Saxon *gor*, 'dirt' (Yorkshire, Oxfordshire)

A Carrion Crow's black beak distinguishes it from a Rook's:
Black Neb – 'neb' means beak (Northumberland) • **Black Crow** (Nottinghamshire)

Names echoing the Hooded Crow's call: **Scald Crow** (Ireland)

From the bird's colour: **Grey Crow** (Scotland, Orkney) • **Grey Back** (Northern England)
Blue-backed Crow (Ireland) • **Dun Crow** (North Yorkshire)

From the bird's black 'hood': **Hoodie/Hoddie** (Morayshire, Perthshire)

CROW COUSINS

Whether you see Carrion Crows or Hooded Crows in your area depends on where you live. Carrion Crows live in England, Wales and southern Scotland, whereas in north-west Scotland, Ireland and the Isle of Man, the familiar crow is the grey-backed Hooded Crow. The two are closely related – they live and breed in different areas, and have different plumage, but where their ranges overlap, they can interbreed. Like Carrion Crows, 'Hoodies' eat carrion. They are less solitary than Carrion Crows, feeding together in fields.

The male crow bows forward as he calls, proclaiming his territory from a prominent perch.

CROW OF DEATH

The Carrion Crow's funereal colour and diet of dead meat have meant it has often been associated with death. Look out for Carrion Crows on roads with wildlife-rich hedges or verges and fast-moving traffic. They can often be seen feeding on roadkill.

Carrion Crows are often the first visitors to a dead animal. Meat-eating birds of prey have sharp hooked beaks for tearing flesh, but Carrion Crows can only reach meat by pecking at a weak point on a dead animal's body. This is often the eye – a gruesome sight that did not endear the Carrion Crow to people in the days before public execution was made illegal.

Seeing, or hearing, a Carrion Crow was often taken as an omen of death. In East Yorkshire, it was believed that a Carrion Crow perching in a churchyard meant a funeral within a week. Personally, I choose to understand that 'an omen of death' can be taken symbolically, as well as literally – it could mean the death of an old way of life or an old pattern of behaviour.

In Ireland, Hooded Crows were said by Irish Travellers to be old Traveller wives, still wearing their grey shawls on their backs. They were thought of as old friends and always brought good luck. The Hooded Crow's habit of feeding on the bodies of the dead on the battlefield linked it to the Irish goddess of war, the Morrígan. In Lady Gregory's 1904 book, *Gods and Fighting Men*, the Morrígan is described as the 'crow of battle'. In old Irish myths, she often takes the form of a crow, inspiring courage in the hearts of warriors and fear in the hearts of enemies. The Morrígan can be understood as an expression of the sovereign goddess of the land, in her role as guardian and protector of people and place. The Morrígan is sister of (and once may even have been identified as) the goddess Ériu, who gave her name to Ireland – Ériu was said to appear as both a beautiful queen and a black crow.

The Scottish Highlands are also home to Hooded Crows. In Scottish folklore, the Cailleach is a divine ancestor, a creator goddess and a weather deity (the Old Woman of Winter) who embodies the crone form of the divine feminine. The Cailleach can appear in the form of a Heron, a Raven or a Hooded Crow.

In nature, death supports birth in a perfect circle, so that life continues. Crows occupy the mysterious, miraculous realm between death and birth, playing a vital role in the cycle of life. They join the ending of death to the beginning of birth to create the circle, keeping the wheel of life turning around.

In prehistoric times, the dead were excarnated (left in the open air), where carrion-eating birds would clean the bones. No wonder carrion-eating crows have been both feared

and revered. In many ancient cultures, crows were totems. It is clear why the crow can be seen as a soul-guide. In a very literal way, the bird carries the dead into the heavens.

CLEVER CROW

In Aesop's fable *The Crow and the Pitcher*, a thirsty crow drops pebbles – one, two, three – into a pot of water, to raise the water high enough to reach. There's truth in the old tale. The crow family is renowned for its intelligence. The size of their brains in relation to their bodies is equal to those of dolphins and primates (which are, proportionally, only slightly smaller than humans' brains). One study gave Rooks a tube of water with food at the bottom. As in the fable, the birds dropped stones into the tube, one by one, to raise the water level until they could reach the food. In other experiments, Rooks worked together to pull a tray of food into their cage, and even crafted a hook from wire to lever a small bucket of food out of a tube. Crows use human tools too – in Finland, where fishing lines are left in holes in the ice, Hooded Crows reel in the lines to eat the fish. Crows can also, as the fable suggests, count – one study tested (and proved) crows' ability to distinguish between numbers up to 30. Pet crows have been taught to imitate human words, using language in ways that suggest understanding (using people's names, for example).

Crows can recognise themselves and individual humans. There are many anecdotal stories of crows bringing gifts for people who have fed or helped them. Gifts reported include a fir cone, a twig, a dried berry, flowers, feathers, coloured glass, bottle caps and beads.

Like squirrels, corvids store food for hungry days ahead. They can remember where they left things, and how long ago. Crows crack open foods with hard shells, like nuts and shellfish, in ingenious ways. To open a mussel, crows fly up high, then drop it onto a rock, cracking open the shell (hence the old Scottish name for an empty sea-urchin shell: crow's cup). At a driving school in Japan, Carrion Crows were seen placing shelled walnuts in front of the wheels of parked cars, so they cracked open when the cars drove over them. The technique was observed by other local crows and the skill spread through the area. Other road-wise Carrion Crows have even learnt how traffic lights work. They wait until the light turns red, then fly down and put the shelled nut on the road. When the light goes green, the Carrion Crow takes off, and watches from a safe distance as the nut gets run over. It even waits for the next red light to collect the contents of the nutshell. Perhaps a Carrion Crow's skill in cracking things open gave us the word 'crowbar'.

There's evidence too that crows hold 'funerals' – groups of crows gather around the body of a dead crow. Wildlife scientist John Marzluff's research on corvids suggests that, 'Just as a playful crow learns and remembers, a mournful or frightened crow also learns [and remembers] important safety lessons.' The crows do not touch the dead body. It is thought the intelligent birds are learning from the death, perhaps about potential dangers to avoid.

And crows play too. Young crows enjoy sliding, just like human children. They slide down sloping roofs, returning to the top to play the game again and again. In Russia, Hooded Crows slide down the golden domes of Russian orthodox churches.

DOUBLE TROUBLE

Unlike sociable Rooks, Carrion Crows are usually seen on their own or in pairs. Pairs work together to take food from other predators, like Herons. One bird hops about in front of the Heron, distracting its attention, whilst the other sneaks off with the fish from behind. Carrion Crows and Hooded Crows will also mob birds of prey, to drive them away from a nest-site.

'THE CROW AND THE SPARROW' A FOLKTALE FROM PAKISTAN

A story that helps us get to know the Carrion Crow, by highlighting its distinctive colour

Once upon a time there was a crow. A hungry crow. In a hole in a tree was a nest. In the nest were three speckled eggs. Sparrows' eggs. The mother Sparrow sat on the eggs to keep them warm, as pleased as can be.

Kraaaw! said the crow. 'Move aside, little Sparrow. I want eggs for breakfast!'

But the mother bird thought fast. 'Mr Crow! Where are your table manners? I've seen you poking about in the rubbish heap. Before you eat, you must wash your beak!'

It was true; the crow had been rummaging in the rubbish heap – it was full of good things to eat. So the crow flew to the river.

'*Kraaw! Kraaw! Kraaw!*' said the crow to the river. '*Water!*

> *Give me water,*
> *to clean my bill,*
> *so that I can eat my fill.*'

'I will give you some water', said the river, 'if you get a pot to put it in.'

So the crow flew to the potter, in the village.

'*Kraaw! Kraaw! Kraaw!*' said the crow to the potter. '*A pot!*

> *Give me a pot,*
> *to carry the water,*
> *to clean my bill,*
> *so that I can eat my fill.*'

'I will make you a pot', said the potter, 'if you get me some clay to make it with.'

So the crow flew to the farmer's field, where the plough was turning up the rich brown earth.

'*Kraaw! Kraaw! Kraaw!*' said the crow to the earth. '*Clay!*

> *Give me some clay,*
> *to make the pot,*
> *to carry the water,*
> *to clean my bill,*
> *so that I can eat my fill.*'

'I will give you some clay', said the earth, 'if you get a spade to dig it with.'

So the crow flew to the forge, where the blacksmith's hammer beat molten metal.

'*Kraaw! Kraaw! Kraaw! A spade!*' said the crow to the blacksmith.

> '*Give me a spade,*
> *to dig the clay,*
> *to make the pot,*
> *to carry the water,*
> *to clean my bill,*
> *so that I can eat my fill.*'

'I will give you a spade', said the blacksmith, 'if you get me fire for my forge.'

So the crow flew to the farmhouse, where the farmer's wife was putting out the scraps on the rubbish heap.

'*Kraaw! Kraaw! Kraaw!*' said the crow to the farmer's wife.

> '*Fire! Give me fire!*
> *to forge the spade,*
> *to dig the clay,*
> *to make the pot,*
> *to carry the water,*
> *to clean my bill,*
> *so that I can eat my fill.*'

The farmer's wife took a burning coal from the hearth. It sizzled in the claws of the tongs, red-hot. 'But how will you carry it?' she asked.

'On my back!' cawed the crow, and he spread his wings out wide. And the farmer's wife put the burning coal upon the crow's back. A sooty, singy smell … Swirling smoke! Blazing flame! Feathers on fire!

'*Kraaw! Kraaw! Kraaw!*' The crow plunged into the river. *Sizzle … fizzle … hiss* … The crow wasn't hurt, but his feathers were burnt black.

So the little brown Sparrow kept her eggs, and hatched three beautiful babies. *Cheep! Cheep! Cheep!*

As for the crow, his feathers are still black, black, black, to this very day.

If the 'Hoodie' is your local crow, rather than the Carrion Crow, this tale can easily be adapted to reflect the distinctive black and grey markings. You could add that Hoodie's ash-grey feathers were created by the ashes of the fire.

THE CROW'S NEST

> *On the first of March,*
> *The crows begin to search.*
> *By the first of April*
> *They are sitting still.*
> *By the first of May,*
> *They've all flown away,*
> *Coming greedy back again*
> *With October's wind and rain.*

NURSERY RHYME

Once paired, Carrion Crows stay together for life, keeping the same patch of ground all year round. Unlike Rooks and Jackdaws, crows nest alone. Hooded Crows nest in trees or on crags, but Carrion Crows usually nest high up in the fork of a tall, solitary tree. This gives them open views in all directions, just like the lookout post at the top of a ship's mast, which is named after the Carrion Crow's creation: the crow's nest.

A pair works together to make their nest – a large, bulky cup (around 45cm wide), made of sticks about the size of a pencil. It is softly lined with wool and fur. Carrion Crows are incredibly adaptable, even nesting on tall buildings. A pair once nested at Heathrow airport, making their nest almost entirely out of scraps of wire. Sometimes, 'nest helpers', non-breeding crows (often last year's young or other relations), help too. An older sibling from a previous year may even stay around to help feed the new babies.

HOW TO DISTINGUISH BETWEEN JACKDAW, ROOK AND CARRION CROW

To spot the difference between these three close cousins, look carefully at each bird's size, colour and beak. Listen closely, too.

The Rook has a high forehead; the Carrion Crow's crown is flattish

The Rook's beak is long and grey; the Carrion Crow's beak is stout and black

The Rook has 'baggy-trouser' leg feathers; the Carrion Crow's legs are neat and sleek

The Jackdaw is smaller than the Carrion Crow and the Rook.

ROOK

CARRION CROW

Rooks walk sedately, with a deliberate stride; Jackdaws walk more quickly, with a sprightly strut

The Jackdaw's grey nape stands out clearly

Its 'tchack' call clearly distinguishes the Jackdaw from the Rook and the Carrion Crow.

JACKDAW

JACKDAW

In silhouette, a Jackdaw's broad neck and chunky bill make its head look short. A Rook's long pointed beak makes its head look long

Jackdaws fly with fast wingbeats; Carrion Crows 'row' across the sky with slow, deliberate wingbeats. A little ditty to help you remember:

*Slow row Crow,
Fast flap Jack!*

ROOK

CARRION CROW

The clearest distinguishing feature between the Rook and Carrion Crow in flight is the Rook's long, pale beak

Though midwinter is behind us, in the British Isles February
can often be the coldest month of the year. The trees are bare.
Wild wolf-winds scour the land. Sometimes, there is snow.

Just as birds are easier to spot in the bare winter branches, individual
birds' voices are easier to distinguish in the wintertime. The cold months
are the best time to begin to learn birdsong, when fewer birds are singing,
and summer visitors have not yet arrived to add their voices to the mix.
It is good to recognise our feathered friends by their voices, since often
we hear a bird before seeing it, or hear it but never see it at all.

Heard even on mild February days, a Great Tit's song is one of the
first of early spring. In this chapter, we get to know the voices and ways
of life of two of our most familiar garden visitors, Great Tit and
Blue Tit, and meet the tiny Long-tailed Tit too.

This chapter also helps us to tune in to the different kinds of
sounds birds make by introducing the basics of 'bird language'.

FEBRUARY

Bird Language

Birds make many different types of sounds as they communicate with each other. Nature Mentor Jon Young teaches that birds have five voices: songs, contact calls, alarms, juvenile begging calls and territorial calls. Here we explore some of the basics.

Songs

Songs are usually sung by male birds, to declare their territory to other males. Singing spaces out the male birds in a particular area, so that each has its own spot, and boundaries between patches are established without any need for physical confrontation. The male bird's song tells the female that he has a place where they can nest and find enough food for a family.

Most often, male birds sing only in spring, to attract a mate to raise a family. Most songs have a sense of pattern – some tuneful element of rhythm or melody. There are exceptions – a House Sparrow's song, for example, is simply a repetitive series of monotone chirps.

Many birds perch on a 'songpost' to sing (such as a particular treetop). See if you can spot the places in your own garden that are regular singing-spots for individual birds.

Contact Calls

Contact calls are sounds that help birds of the same species to keep in touch with each other, and keep the flock together. They tell companions, 'Here I am … where are you? I'm ok … Are you ok?' Contact calls are often short and soft, rhythmic and repeated, as birds stay in touch whilst they move around. Generally, smaller birds keeping in contact as they forage in bushes use quieter contact calls. Larger birds in flight may use louder contact calls; the honking of geese and the *tchack*-ing of Jackdaws flying overhead are both loud and carrying contact calls.

Begging Calls

Begging calls are made by young birds asking their parents for food. Whilst the young birds are in the nest, the calls come from one spot. Begging calls are often incessant, and have an insistent tone – think of the sound of a human baby crying to be fed.

Territorial Calls

If you hear noisy, intense calls coming from one spot, especially in spring, it could be a quarrel between two neighbouring birds of the same species, as two males come to an agreement over territory boundaries.

Territorial calls, begging calls, contact calls and songs are what Jon Young calls 'baseline' behaviours, meaning birds' 'normal' behaviour, when there is no threat from a predator. In contrast, the fifth type of call is made in direct response to a possible threat (below).

Alarm Calls

Alarm calls warn every species (bird and mammal) of danger. It can be anything from a big noise, such as a Blackbird's urgent *pink-pink-pink-pink-pink-pink-pink*, to a complete silence, as when a Sparrowhawk flies by. The tone of the sound (or the held-breath silence) expresses the feeling of alarm.

Birds have different alarm calls to signal the presence of ground threats and threats from the sky. The soft *chook ... chook-chook-chook* of the Blackbird warns youngsters of a ground predator (a cat, perhaps, or a human), so they keep quiet and stay still. The Blackbird uses a different sound, a high-pitched *seeee*, to signal danger from above. A similar sound is used as a 'hawk alarm' by many different species of birds. If you hear it, look to the skies – you may spot a bird of prey.

Some birds even give specific calls for specific species. The Jay is a skilled mimic and imitates the voice of the predator, giving a chittering call for a squirrel and a shrill *ki-ki-ki* for a Kestrel. The Great Tit, too, uses different alarm calls to warn of different predators; *jar* sounds mean snakes and *chicka* sounds mean crows. The Great Tit alarm call for a marten uses the same sounds as an alarm call for a crow, but in different combinations – just as we use the same sounds in different combinations to create different words.

Mobbing Calls

An alarm call usually causes the bird that hears it to escape by fleeing or hiding. In terms of 'fight or flight', it could be thought of as a call to 'flight'. A mobbing call is a special kind of alarm call that requests a different response. The mobbing call attracts other birds (sometimes including other species) to join in with directing alarm calls at a predator, such as a cat or sleeping owl, with the intention of driving the predator away. It could be thought of as a call to 'fight'.

'THE LAD WHO LEARNT THE LANGUAGE OF THE BIRDS' A FOLKTALE FROM FINLAND

A story that introduces the concept of bird language

Once there was a king who loved learning. When he became a father, he sent for the best teachers in the world to teach his daughter. The princess spent every day in the schoolroom. She learnt every language under the sun. She loved the way each language was a new music to her ears, a new way of seeing the world and understanding others. She learnt Finnish and Danish, Swedish and Norwegian, Greenlandic and Icelandic, Russian and Arabic, English and Estonian. And all the other languages too. Sometimes, she sat at the window, looking out over the forest, wondering about the wide green world beyond.

Well, when it was time for the princess to marry, her father proclaimed, 'Only a man who can match my daughter's wisdom is worthy of her hand. Any man who wishes to court my daughter must prove to me that he can speak a language she herself does not know. But beware! If any man dares to woo my daughter who has not proved his wisdom, he will be flung into the sea!'

Well, many men tried. Rich men. Grand men. Important men. But, *SPLASH! SPLASH! SPLASH!* Not one of them could speak a language the princess herself had not already mastered.

There was one young man, a shepherd lad, who wished to court the princess. He wasn't rich and he wasn't grand. He hadn't been to school. But he loved to learn. Nature was his teacher. He loved to wander and wonder in the forest, looking at the trees and the plants and the flowers, and listening to the animals and the birds. He knew where the Robin made her nest and where the Blackbird sang his song. He'd seen the king's daughter gazing out of the castle window, and he had smiled up at her, and she had smiled back down at him.

So off he went through the forest, to the king's castle. He hadn't gone far when he heard a squeaky peepy sound. He looked up at the old silver birch, and saw that the noise was coming from a little hole in the trunk. He sat down quietly to watch. Soon, he saw a mother bird, flying to the nest, with her beak full of food. The baby bird squeaked and peeped with all its might! Off she flew again. The boy sat still, and watched. Soon the father bird flew in, with his beak full of food. *Squeak-squeak-squeak!* The boy smiled. He sat watching for a while. Then off he went on his way.

He hadn't gone much further when he found a feather. It was patterned with warm brown and rich cream, and, looking closely, he saw a tiny fringe along the feather's edge. He stroked the feather against his cheek. So soft. He swished it beside his ear. So silent. An owl feather. He put the feather into his leather pouch and went on his way.

Before long he came to the castle. He looked up at the great stone walls and tall round towers. On the edge of a turret, there perched a grey-brown bird, chirping its noisy notes over and over again. A Sparrow, singing its simple song. The boy smiled. He knocked at the great door.

Inside, the king and his daughter were seated on golden thrones. 'Well?' said the king, looking at the boy's humble clothes. 'The sea is very cold at this time of year! Are you sure you want to risk my challenge?'

But the boy planted his feet firmly on the ground. 'Yes,' he said. He looked up at the princess. 'I'm sure.'

The boy called the princess to stand beside him by the window. 'Listen,' he said.

The princess tilted her ear, smiling at this new game. She was good at listening, but she had never listened so closely to the forest before. From the trees came a squeaky, peeping sound: *wi-wi-wi-wi-wi*. It was very insistent.

'Listen,' said the shepherd boy, 'and tell me if you understand …'

The princess shook her head. 'I don't,' she said. 'That is a language I've never learnt. What does it mean?'

'It means: "Mummy! Daddy! Feed me! Feed me!" It's a baby woodpecker. If you like, I can show you its nest …'

At the thought of seeing a real live woodpecker, a little chick all fuzzy round the edges with baby feathers, the princess's eyes shone.

'Now', said the shepherd boy, 'look there, on the corner of the roof. Can you hear that chirping?'

The princess tilted her ear and looked. She nodded.

'Can you tell me what that Sparrow is saying?'

Again, the princess shook her head. 'He's saying it at the top of his voice!' she said. 'Is it something very important?'

'It is,' said the shepherd boy. 'He's saying: I have a good place here. There's a hole in the stone just right for a nest. Dear sparrow-maiden, do come and join me!'

The princess was enchanted. But the king was furious. 'Daughter!' he cried, 'You've had the best teachers in all the world, and yet this country boy makes fools of us both.'

'I'm sorry, Father,' said the princess. 'Perhaps I'm not so clever, after all.'

'My king,' said the shepherd boy, 'your daughter is no fool. She is clever enough to know that she doesn't know it all, and to ask for help. She asks questions and she follows her curiosity. She understands that there is always more to learn – as I see it, that is true wisdom.'

At the boy's words, the king's face softened. He nodded.

Outside, in the forest, dusk was beginning to fall. The shepherd lad and the princess stood together by the open window, and breathed in the soft scents of the evening air. *Too-whit … Too-whit …* The wavering hoot of an owl floated over the trees.

'I wonder what the owl is saying?' said the princess, leaning on the windowsill and looking out into the night.

The shepherd boy leant beside her. 'It's a Tawny Owl, calling to her mate,' he said. 'She's saying: "I'm here … where are you-ooooo …"'

As they listened, they heard a wavering *too-whooooo*. 'And her partner is answering: "I'm here, my dear …"'

The princess turned to the boy, smiling. What a wonder to understand the language of the birds!

'If you like,' said the shepherd boy, 'I'll take you into the forest, and we can listen and learn together.'

'Oh, yes please!' said the princess. 'Father, may I?'

The old king looked at the shepherd boy with a new respect in his eyes. 'You do know a language my daughter has not yet learnt. It took courage to take the risk. And you have risen to my challenge. If my daughter wishes it, you may court her.'

And so, the princess and the shepherd boy spent many happy days wandering the forest together, looking and listening, learning, and falling in love.

GREAT TIT

Parus major

White cheeks

Black cap and bib

Thick black stripe; female has fainter belly-stripe

Yellow belly

Juveniles have yellow faces.

VOICE

A Great Tit's most frequent and familiar spring sound is a loud two-note song: *tea-cher, tea-cher, tea-cher, tea-cher*. The song has the see-saw rhythm and metallic tone of a saw moving back and forth. The Great Tit also has a sharp *zinc* call.

SIZE

14cm

WHERE

Woods, parks, hedgerows, gardens.

NEST

Female makes nest in hole in tree.

EATS

Insects, seeds, fruit.

GREAT TIT – THE OLD NAMES

'Great' Tit is the biggest European tit. 'Tit' is short for 'titmouse' – old Icelandic *tittr* means small thing, Anglo-Saxon *Mûse* means small creature

From the Great Tit's black 'cap':
Black Cap/Black-capped Lolly (Northamptonshire, Yorkshire) • **Black-headed Bob (Devon)**

From the Great Tit's saw-like song:
Saw Sharpener (Roxburgh) • **Sharp Saw (Norfolk)** • **Joe Ben** – echoing the song's two-note rhythm (Suffolk)

The bird's white patch was thought eye-like:
Ox-Eye (Midlands, Shropshire, Yorkshire, Ireland) • **Big Ox-Eye (Forfar, East Lothian, Roxburgh)**

The Great Tit protects its nest, and food, boldly:
Guerrero – 'warrior' (or 'brawler') (Spain) • **Heckymal** – from Saxon *hick*, peck (Dartmoor)

THE GREAT TIT'S SONGS

Great Tit's simple see-saw song (likened to a squeaky bicycle pump) is one of the most familiar sounds of early spring. Yet the Great Tit also has an amazing variety of different songs, perhaps the most of any British bird. If you hear a bird you can't recognise in the woods, it's probably a Great Tit.

THE GREAT TIT'S NEST

As early as February, you may see a male Great Tit courting a female, attracting her attention by sticking out his chest and fluffing his feathers, and lifting his head to show off his belly stripe.

In the weeks before laying her eggs, the female increases her weight by at least half, a feat that cannot be achieved without her mate's help. You may be lucky enough to see a Great Tit pair side by side on a branch, the male offering his mate a beakful of caterpillars, which she receives with open mouth and fluttering wings (a gesture similar to a baby bird begging for food). This 'courtship feeding' continues whilst the female makes the nest, with her mate bringing up to 50 deliveries a day. The male also keeps watch for danger, freeing the female from the need to constantly look around, so she can spend more time feeding.

Great Tits usually only have one large brood each year. They literally have all their eggs in one basket. The female sits on the eggs to keep them warm, and during this time the male continues to feed her. She turns the eggs with her beak so they're warmed evenly, and takes regular breaks to forage and stretch her wings.

Once the eggs have hatched, around mid-April, the Great Tits bring insects to their brood – visiting the nest hundreds of times a day.

The young are fed mainly on the caterpillars of moths – their hatching is perfectly timed with the weeks when the caterpillars are most abundant.

PROTECTIVE POWER

Great Tits defend their nests with hearty courage. In a nest box, the mother bird will continue to sit, even if the lid is lifted for a peep inside.

If a predator, such as a weasel, comes near the nest-hole, the female Great Tit will not flee. Instead she lunges forward, shaking her wings and tail feathers, and hisses like a snake. The sound of a snake is often enough to frighten an intruder away.

MOBBING

To the ears of a Great Tit, mobbing calls are lower in frequency than alarm calls, carrying over longer distances to summon birds from all around to drive a predator away. As well as different alarm calls, Great Tits also use different mobbing calls for different predators – the call to mob a Tawny Owl is different to the call to mob a Sparrowhawk.

Male Great Tits are especially willing to risk their own safety to mob a predator. If a Great Tit discovers a sleeping owl, it scolds loudly, dodging back and forth, calling other birds. In the branches, a Robin *tic-tics*. A Wren shrills. Owls rely on surprise. Without it, they cannot catch their prey. Before long, the owl flies away. Mobbing disrupts and disturbs a predator. By moving an owl on, a bird could be removing it from its own territory, nest, eggs or family. Mobbing might also demonstrate to a predator that a certain bird is too agile and alert to bother trying to catch. It might teach young birds about local dangers. Mobbing an owl also disturbs its sleep, perhaps making it a less efficient hunter that night.

WATCH GARDEN GREAT TITS FEED

At the garden feeder, Great Tits can also be a fierce presence. A Great Tit will budge smaller Blue Tits away with a single squawk, and see off other Great Tits that come too close. You might see a Great Tit gesturing 'back off!' by:

- Opening its wings, perhaps whilst hissing or scolding.
- Lowering its head, pointing it towards the 'intruder'.
- Raising its chin high, showing off the broad black stripe on its belly.

Another Great Tit might respond by signalling its submission. It might ruffle its plumage and lean forwards (minimising the effect of its own black stripe). It might fly off, to the far side of the feeder. These visual and verbal signals allow the birds to settle the matter without a fight, so neither bird is harmed.

Though you may occasionally see Great Tits and Blue Tits disagreeing over food on garden feeders, in the wild they coexist in harmony. Bird tables and hanging feeders bring birds into unnaturally close contact. In their natural woodland habitats, though the two birds often eat much the same food, they feed in different places. The light little Blue Tit feeds in the tops of trees, hanging on the very tips of the thinnest twigs, amongst the buds. The Great Tit is too heavy to forage in the fine branches high in the canopy; it feeds in the thicker branches, lower down, and on the trunk. Great Tits also feed on the ground, looking through the leaf litter for seeds and nuts. The Great Tit can hack open a hazelnut with its strong beak, and (like the squirrel) hides nuts in the autumn. In winter, watch the birds searching the woodland floor to find them again. Long-tailed Tits, who might also join a mixed flock in winter, are so tiny they're not strong enough to flick fallen leaves aside. Their long tails, too, are cumbersome on the ground. They feed amidst the thin twigs, where the Great Tit is too heavy to perch.

Great Tits are ingenious feeders – they've even been known to use pine needles as tools to extract larvae from holes in trees.

If you would happy company win,
Dangle a palm-nut from a tree,
Idly in green to sway and spin,
Its snow-pulped kernel for bait; and see,
A nimble titmouse enter in.

FROM 'THE TITMOUSE', WALTER DE LA MARE

A TIT FLOCK

Through spring and summer, all tits feed on insects. In autumn and winter, when bugs and caterpillars are less plentiful, families of Great Tits flock together with Blue Tit families to forage. In woodlands, they may be joined by Coal Tits. You can recognise this bird's sweet song, which sounds like a Great Tit in miniature; if a Great Tit's song sounds like a bicycle pump, a Coal Tit's song is a toy bicycle pump.

Go for a walk in a deciduous wood between August and March, and hear the foraging party calling noisily to each other as they roam through trees and hedges. They may even visit your local park or garden. Use binoculars to spot which birds are part of the flock.

GREAT TIT

BLUE TIT

Cyanistes caeruleus

Blue cap

Dark line through eye, like a 'bandit mask'

Narrow blue-black chin

Yellow breast

Juveniles have yellow faces, but can be identified by their distinctive 'bandit mask'.

Blue wings

VOICE
Calls with two or three whistled notes: *tsee-tsee-tsee*. A Blue Tit's song starts with the call and ends with a pretty tinkle, like a tiny bell being shaken quickly back and forth: *tsee-tsee-tsee-tissississi*. Alarm call is a scolding *churr*.

SIZE
12cm

WHERE
Woodlands, farms, gardens, towns.

NEST
Female makes moss-nest in tree-hole.

EATS
Insects, seeds.

BLUE TIT – THE OLD NAMES

Tom Tit – short for 'Tom titmouse', an English folk-name • Titw Tomos Ias (Wales)

From the Blue Tit's blue 'cap':
Blue Cap/Blue Bonnet (Shropshire, West Riding, Scotland) • Blue Ox-eye (Forfar)
Blue Yaup (Scotland) • Blue Spick (North Devon)

Nun – the white feathers encircling the blue cap were thought similar to a nun's white headband

From the way the Blue Tit defends its nest:
Hickmall/Hackmal/Heckymal/Hagmal/Titmal (Devon, Cornwall) – from Anglo-Saxon
for Blue Tit, *Hicemase*, 'hick' (and derivatives) refers to 'hacking' pecks • Tom Bitethumb
Billy Biter (Shropshire, North Riding, Sussex)

COLOURFUL...

Like most birds, Blue Tits can see ultraviolet (UV) light – in UV light, the bird's blue cap glows bright. One of the ways a male Blue Tit shows off to a female is by raising his blue head feathers into a 'crest' (he also uses this gesture to see off other males). Studies show that both males and females choose mates with brighter caps.

The yellowness of a Blue Tit's breast depends upon the number of yellow-green caterpillars the bird has eaten. In males, a brighter breast is more attractive to females, as it shows her potential mate is good at finding food – an essential skill in bringing up a big brood.

...AND CHARMING

In the wild, Blue Tits often hang upside down to search the underside of leaves for caterpillars. In our gardens, their acrobatic antics make Blue Tits an entertaining sight.

Blue Tits enjoy peanuts. On a winter's day, making a string of nuts is a satisfying activity for children, and watching a Blue Tit as it feeds is a sight to cheer young and old. Simply provide children with a blunt needle, a length of wool and a bowl of monkey nuts – the needle can be passed through the narrow part of the shells between the two peanuts.

APPLE BLOSSOM

If you are lucky enough to have an apple tree in your garden, look out for Blue Tits in the branches; a delightful sight amidst the rosy blossom in spring. The birds search for the grubs that feed in the buds, as poet John Clare says:

The bluecap toothless in its glee
Picking the flies from orchard apple tree

Blue Tit chicks can eat a hundred caterpillars a day, and there can be up to 14 chicks in a single brood; that's a lot of caterpillars! When an accident left him unable to travel, writer H. J. Massingham valued the birds in his own garden all the more. He recorded watching Blue Tit parents flit between apple tree and nest – estimating the brood received 1,500 meals a day.

MILK BOTTLE TOPS

Blue Tits are experts at pulling bark off trees to find insects, and in the early twentieth century used this skill to pick off tiny pieces from the foil lids of doorstep milk bottles, tossing them aside until they eventually broke through to reach the cream. The behaviour was first recorded in Hampshire in 1921. Other local Blue Tits saw the attention the bottles were getting, and investigated the lids for themselves, soon working out how to open them. So the skill spread. In the 1960s and 1970s there was barely a doorstep in Britain that Blue Tits hadn't borrowed milk from (and Great Tits had also learnt the skill from their cousins). Sometimes Blue Tits followed milk floats down the road, or waited for delivery time, and swooped down when the milk was put on the step.

EGGS IN ONE BASKET

Like Great Tits, Blue Tits usually have just one brood each year, and time the hatching to coincide with the greatest abundance of caterpillars. In early summer, the caterpillars of the winter moth feast on young oak leaves, and Blue Tits feast on the caterpillars. The oak is a Blue Tit's favourite nest tree; the bird even prefers nest boxes in oak woods.

The female lays one egg each day, until she has up to 12 eggs (the total weight of her clutch is the same as her own body weight). She only begins to sit on the eggs to warm them once the clutch is complete, so if the female has to leave the nest for a while, she covers the eggs with some snug nest lining, like a mother tucking a baby under a warm blanket. Make a nest-material ball (instructions on page 55) and watch Blue Tits and Great Tits gather soft wool to keep their little ones warm.

By midsummer, Blue Tits that nested in garden boxes in spring can be seen attending their noisy brood. Listen for the incessant calls of fledgling Blue Tits, as they follow their parents around the garden.

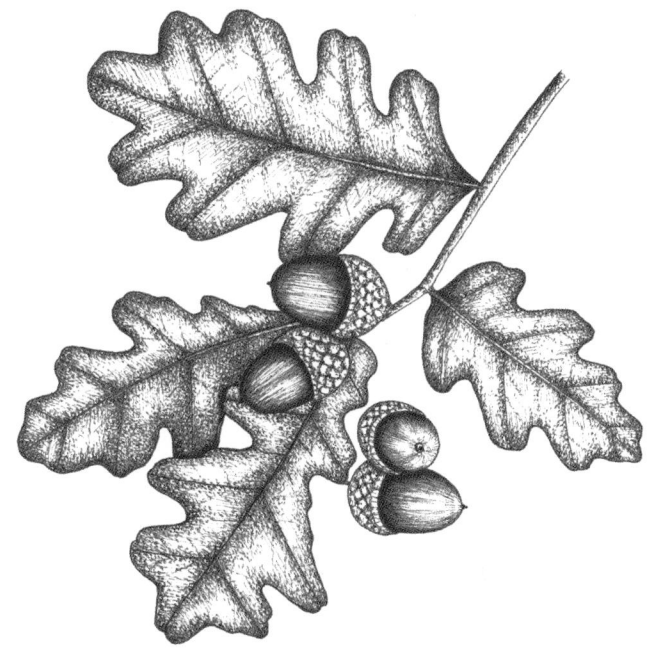

PROTECTIVE PARENTS

Small as they are, Blue Tits, like their cousins Great Tits, stay on the nest even if disturbed, and hiss like a snake to scare off predators, attacking anything that comes through the nest-hole. Parents also scold loudly if a thoughtless human blocks their direct line of flight to the nest. In the days before egg collecting was made illegal, schoolchildren soon learnt that a female sitting on eggs will deter curious fingers with a peck.

LEARN BLUE TIT'S SONG

Conservationist Lucy Lapwing writes the whistled three-note call of the Blue Tit as *tea-tea-tea*, giving a visual memory aid of ☕☕☕ to make it easy to remember. For the Blue Tit's pretty song, which begins with the call and ends with a bell-like tinkle, Lucy writes *tea-tea-tea, lily-lily-lily*: ☕☕☕ 🌸🌸🌸.

LONG-TAILED TIT

Aegithalos caudatus

Long tail

Belly has a
hint of pink

Juveniles have
shorter tails and
no pink.

VOICE
The Long-tailed Tit's call is a
thin, high triple squeak: *zi-zi-zi*.
A flock of Long-tailed Tits calls
to each other continuously as they
cross from one tree to another. The
song – an elaborate version of the
calls – is not often heard.

SIZE
12–16cm, most
of which is tail
(the bird's body
measures only
5–6cm)

WHERE
Woodlands,
hedgerows,
parks, gardens.

NEST
Rounded nest in
spiny shrub.

EATS
Insects,
occasionally
seeds.

LONG-TAILED TIT – THE OLD NAMES

From the Long-tailed Tit's long tail:
Kitty Longtail • Long Tom • Long-tailed Mag/Long Pod (Midlands)
Long-tailed Chitterling – 'chitter' evoking the birds' high-pitched chatter
Long-tailed Pie – from the 'pied' plumage (Cheshire)

From its tiny size: **Millithrum** – meaning 'miller's thumb' (akin to 'Tom Thumb')
Creak-mouse – 'creak' evoking the bird's squeak (Gloucestershire)

From its feathers, which look fluffy as a muffler:
Long-tailed Muffin – 'muffin' is also a term of affection • **Long-tailed Mufflin**
Mumruffin (Worcestershire, Shropshire) • **Ragamuffin** • **Fuffit** (East Lothian)

From its round nest:
Nimble Tailor • Jack-in-a-bottle/Bottle Tom • Bottle Tit (West Riding, Berkshire,
Buckinghamshire, Shropshire) • **Hedge Jug • Oven Builder** (Stirling) • **Bush Oven** (Norfolk)
Barrel Tom • Bum Barrel (Nottinghamshire) • **Pudding Bag** (Norfolk)

From the feathers poked inside the nest:
Feather Poke • Poke Pudding (Gloucestershire, Shropshire) • **Puddney Poke** (Norfolk)

LITTLE BIRD, LONG TAIL

A Long-tailed Tit's tail is bigger than its body. Its shape makes it easy to recognise – it looks like a ball of feathers with a tail stuck on. With its dainty size, fluffy feathers, pretty colours and ball-and-stick shape, the Long-tailed Tit is sweet as a lollipop. The long tail helps the bird balance, acting as a counterweight during acrobatic feeding.

> *Bum-barrels twit on bush and tree*
> *Scarse bigger then a bumble bee*
> *And in a white thorn's leafy rest*
> *It builds its curious pudding-nest*
> *Wi hole beside as if a mouse*
> *Had built the little barrel house.*

FROM 'THE SHEPHERD'S CALENDAR', JOHN CLARE

THE WONDER NEST

The birds choose a nest-site low in a thorny bush. Both parent birds build the nest, a master-craft creation that begins in March and can take three weeks to complete.

First the birds collect moss from trees and logs and drop it onto a thorny branch until it lodges to form a foundation. They gather spider's silk to bind the base to the branch. Then they sit on the platform and, turning slowly around, build a cup of moss and cobwebs around themselves. Both birds spend several days collecting lichen (as many as 4,000 flakes) from the trees, to weave into the outside of the nest for camouflage. When the moss cup is higher than their own heads, the tall walls lean in to create a roof, and are woven together.

The pair now focus on finding feathers, collecting up to 2,000 over several days. The feathers left over when a pigeon is eaten by a fox or hawk are a valuable find for Long-tailed Tits. The feathers are poked inside to line the nest – making a fluffy feather bed.

The finished nest is a cosy oval ball and the parents have to fold their tails back to fit inside. Thanks to the supple spider's silk, the nest can stretch, meaning that as the baby birds inside grow, the nest grows with them.

Once the nest is complete, the female begins to lay her eggs. Every morning she lays one egg, each not much bigger than a peanut. She lays around 10 eggs (which together weigh more than she does) and incubates them for around two weeks. Her mate may bring her a mouthful of insects now and then, and she also leaves the nest regularly to forage. The feather bed keeps the eggs warm whilst she feeds. As the chicks grow, the space inside the nest shrinks. Watch out for a female with a curled-up tail – the result of being in her very snug nest.

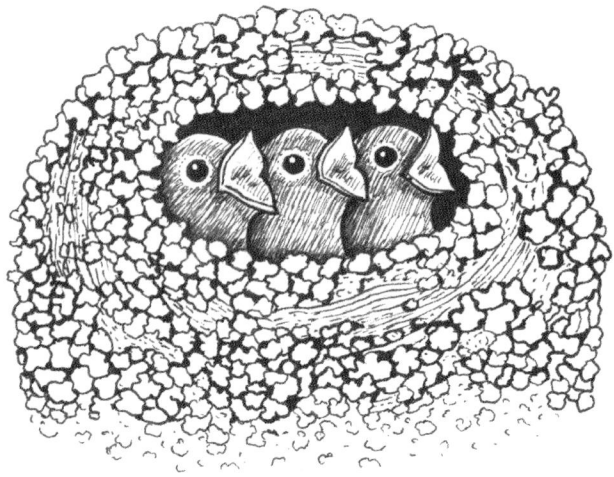

A FAMILY FLOCK

When the chicks leave the nest, the parents stay with them, forming a family flock. From summer onwards, they travel and feed together, roving through hedgerows, shrubs and trees.

Over winter, the male youngsters stay with their parents in the family flock, and the female youngsters leave to join a neighbouring family group. At the same time, unrelated females from neighbouring family parties join the flock. All the birds in the flock defend a single group territory.

When spring comes around, the flock splits and the group territory is split between them – so the newly paired young birds each have their own patch to nest in.

Magpies, weasels and other predators all eat the eggs and young of Long-tailed Tits – more than half of Long-tailed Tits' nests are lost in this way. If this happens birds may nest again, if there's time in the season, but if not, they often offer help at a nearby nest instead. Most often a male bird will help a sibling (since females are less likely to have relatives nearby, because they flew further from home in their first winter). The 'kindly uncle' helps build the nest and raise his nieces and nephews. The brood's chances of survival are higher with more carers, and the helper becomes part of the brother's flock, which increases his own chances of surviving the cold months.

Professor Ben Hatchwell, from the University of Sheffield, studied a population of Long-tailed Tits for over a decade. As he observes, the cooperation of Long-tailed Tits 'appears to contradict Darwin's theory [of natural selection]'. Darwin's concept of 'survival of the fittest' has historically been interpreted as implying that 'selfish' competition is the natural way of the world. Yet Darwin also wrote of cooperative behaviours in many species. Like squirrel-and-oak, Blackbird-and-bramble, foxglove-and-bumblebee, Long-tailed Tits give us

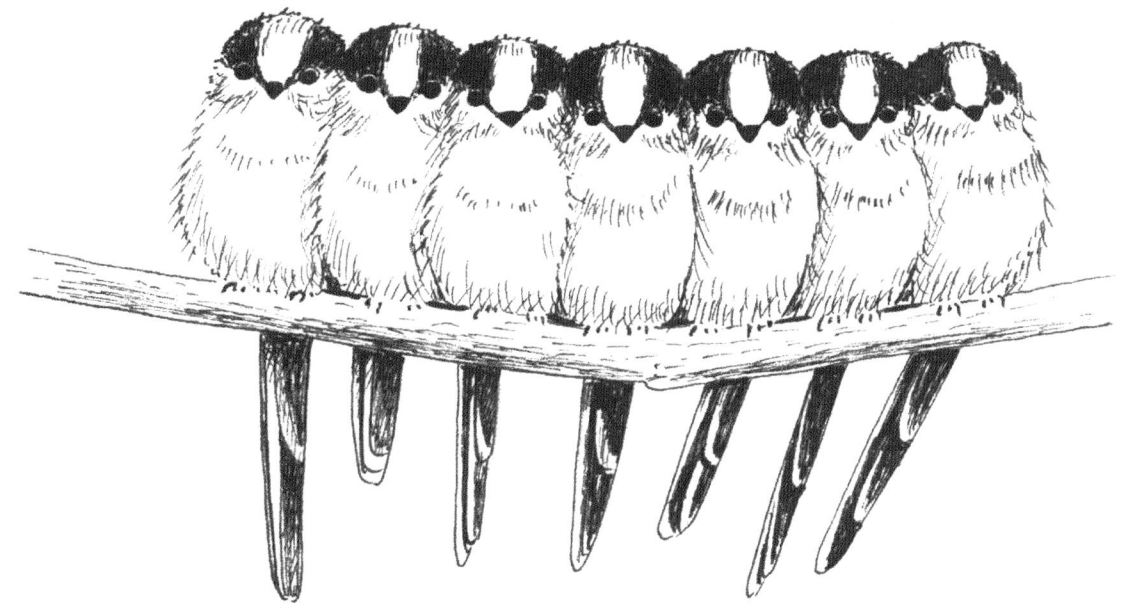

Most birds that roost communally – such as Starlings – sleep near to each other but not in bodily contact. Long-tailed Tits snuggle close together so they're warmed by body heat on both sides. Older birds get the privilege of sleeping in the centre, the warmest spot.

a touching example of how, in the natural world, living beings cooperate in ways that benefit all involved – relationships that weave together the web of life.

Indeed, Professor Hatchwell's research showed that 23 per cent of birds helped those who were unrelated, but known through social ties (i.e. friends). A Long-tailed Tit might help an 'old flame' (from a previous breeding season), a child, a parent, an aunt, even a grandparent. One brood can be cared for by up to eight helpers, and even after the young leave the nest, the close relationships between parents and helpers continues. Long-tailed Tits offer us plenty of inspiration for living well: they have strong family ties and lasting friendships and work together to help each other thrive.

SLEEPING SIDE BY SIDE

Because they're small, Long-tailed Tits lose heat fast, and are particularly vulnerable to the cold. Their fluffy feathers hold a layer of warmer air close to their bodies. Still, they need more heat to survive cold nights. When chicks are young, both adults sleep in the nest with them, snuggled together in their high-tog feather bed. After the young have flown the nest, an extended family of Long-tailed Tits (a mother and father, their fledglings and a number of aunts and uncles related to the family on the father's side) will perch side by side. Most birds that roost communally sleep near to each other without touching. Long-tailed Tits snuggle close together, warmed by the whole family.

'THE TOMTIT (BLUE TIT) AND THE BEAR' A FOLKTALE FROM GERMANY

A story that introduces the concept of mobbing and celebrates the tit family's protective parenting

In a wood, in a tree, in a hole, was a nest. A tomtit nest, made of soft moss and warm wool. A cosy hole, stuffed with softness.

And in the nest – *cheep, cheep, cheep* – were six little chicks. Hungry little chicks. Two little tomtits, the mother and father, went back and forth, back and forth, with beaks full of fat caterpillars.

Well, who else lives in the forest? Someone shaggy and shuffling, big and brown ... Bear! Bear was looking for good things to eat. Aha! A hole! Sometimes, a hole in a tree means bees, and bees mean ... honey!

The bear stretched up onto his toes. He poked his nose into the hole.

Inside, Mother Bird was sitting on her nestlings, to keep them warm. When she saw Bear's nose poking into her home: *Hsssss*! She gave a loud hiss, like a scary snake.

Now, a snake would scare off a mouse looking for a house. But it didn't scare Bear. Bear had tough skin and thick fur – he wasn't scared of a snake. He reached up his paw and poked his claws into the hole. Mother Bird wasn't having that! *Hack*! She pecked his paw with all her might.

'Ow!' cried Bear, backing away.

The baby birds were all upset – they cried and cried. Finding food nearby, Father Bird heard their calls. Like a winged warrior, he flew to help.

Mother Bird flew out to join him. All the fierce force of their love coursed through their tiny bodies.

'No!' they scolded. 'No, no, no!' The little birds dodged back and forth around the bear, calling and calling, as he shambled and swiped.

The other birds heard their calls, and came to help. Robin came shouting, 'Tic! Tic-tic!' Wren shrilled, 'Teck-teck-teck!'

Bear covered his ears with his hands. A storm of birds dive-bombed Bear's back, as he turned tail and ran. 'I'll be back!' he roared, 'and I'll bring all *my* friends!'

'Then we'll fight!' squawked Father Bird. 'Tomorrow, at first light ...'

Off Bear stomped. With a mighty roar, he summoned all the furry four-legged folk of the forest to help him: wolf and weasel, stag and stoat.

But the clever little tomtits summoned all the winged ones, the flying creatures of the air, and they sent the smallest of them all – Gnat – to listen to the animals' plan.

Nnnnnnn! Gnat landed on Bear's shoulder, and listened. Bear was talking to Fox. 'Fox,' he was saying, 'you are the most cunning and crafty of us all. So you shall lead us in battle. We will all follow your orders.'

'Good,' said Fox. 'I'll lead the charge. My tail is as good as a red flag. When you see me lift my tail up, charge! But if you see me lower my tail, run away!'

The teeny gnat told the birds all he had heard. And the birds made a plan.

Next morning, at first light, the fox led the animals to the clearing with his tail held high. The beasts charged. A thunder of pounding paws, stamping and tramping, the ground shuddering ...

And the air thrumming and humming with wings. 'Now!' shouted Father Bird. And the hornet zoomed right up to Fox's tail. *Bzzzzzz. Sting!*

'Yow!' Fox's tail swung down as he hopped and howled. 'Ow! Ow! OW!'

All the animals saw the sign – clear as a red flag. 'Run away!'

They went pelting away, bounding and bawling, skittering and skedaddling, away into the forest.

'Good!' said Mother Bird. 'That's the end of that!'

But it wasn't quite the end, because Bear sat brooding in his lair. He sulked and he simmered. He stewed and he chewed over what had happened. His bear brain worked very slowly. He needed time to think things through. He sat mulling and musing.

He thought of his own mate, the great she-bear, and how fierce she could be, when there were cubs to care for. Even fiercer than he was himself! He chuckled. Good old Mama Bear. He thought of the tomtits, so small, but so bold, and he smiled.

After a while, the big old bear stumped back to the tomtits' nest. He didn't poke his nose into the hole. He kept his distance, so as not to frighten the baby birds. They were sitting on a branch, fat fluffy fledglings with fuzzy feathers and yellow beaks. 'I'm sorry!' the bear called to the tomtits. 'I won't disturb you again.'

And, true to his word, he never did.

BUILD A BIRD BOX

In February, many species of birds are house hunting. In the countryside, both the Great Tit and the Blue Tit nest in holes in trees – so both appreciate a garden nest box. You can buy a ready-made box from the RSPB, or building a bird box is a wonderful project for an adult and child to work on together.

You will need:

- Plank of Forest Stewardship Council (FSC) wood, about 150mm wide and 15mm thick
- Galvanised nails/screws
- Waterproof strip of leather/roofing felt/rubber (e.g. a piece of old bicycle inner tube)
- Natural oil
- Pencil, tape measure
- Saw
- Drill
- Hole saw/cutter for making 25mm/ 28mm holes
- Catch
- Wire
- Rawplugs
- Ladder

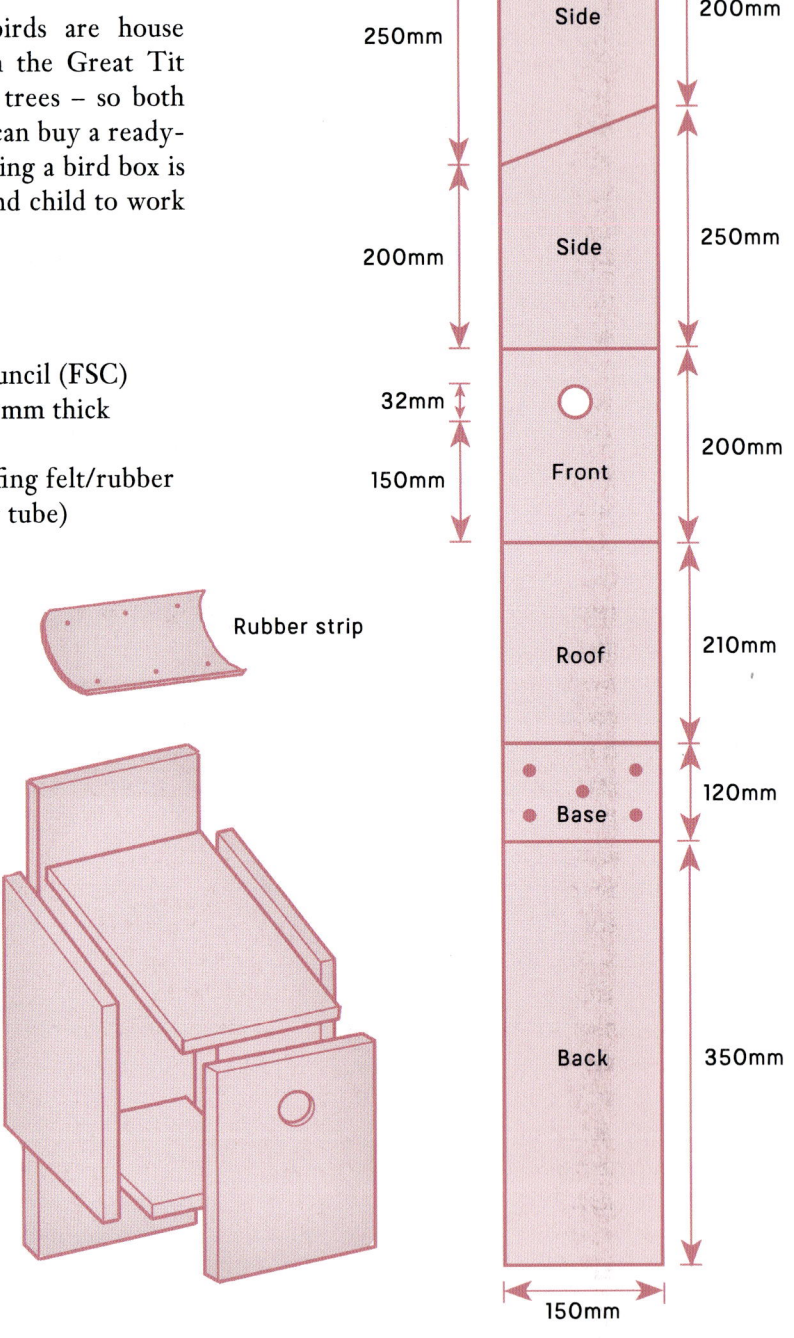

Rubber strip

1 Measure and mark your plank of wood, as shown. Make the inside of the box at least 100mm square. Make the bottom of the entrance hole at least 125mm from the floor (so a cat's paw can't reach in to scoop out young birds). Cut the plank into the pieces shown.

2 To help young birds clamber out of the hole, score the wood on the inside of the front 'wall' piece, creating a rough surface (which is easier for young birds to grip). Make small drainage holes in the 'base' piece, to help stop the box getting damp inside.

3 Choose the size of entrance hole you want: 25mm suits Coal Tits, Marsh Tits and Blue Tits; 28mm suits Great Tits. Keeping the hole size to 28mm or less stops larger birds such as Starlings (who could take over a tit's nest) from coming in. Use a hole saw or a cutter to create the doorway.

4 Use galvanised nails or screws to fix the pieces of wood (except the roof) together. The sides, back and front wrap around the base, as shown.

5 Use the waterproof strip to make a hinge between the top edge of the roof and the back board. Use a catch to keep the lid closed, so you can open it again to clean out the box in autumn.

6 Boxes made of hardwood (like oak or beech) withstand the weather well, lasting for up to 15 years. Boxes made of softwood (like pine) can be treated with preservatives to help them last longer. The best options are 100 per cent natural, such as raw linseed oil or tung oil. You could also use a water-based wood preservative. Apply only to the outside of the box (not around the entrance hole). Make sure children only use preservatives with adult supervision. Allow the box to air dry thoroughly, so harmful fumes can evaporate.

7 Choose the location. Unless there are trees or buildings to shade the box during the day, face the box between north and east, to avoid strong sunlight and the wettest winds. Make sure the birds have a clear flight path to the nest without any clutter directly in front of the entrance. Consider the area below the box site – a soft landing for new fledglings and some shrubby cover are ideal.

8 Put your box up, 2–4m high, on a tree or wall. To attach the box to a wall, drill guide holes in the back plate at the top and bottom of the box. Use a ladder and fix the box to the wall using screws and rawplugs. If fixing your nest box to a tree, rather than using nails (which may damage the tree) attach it with a wire around the trunk or a branch. Use a piece of hose or section of car tyre around the wire to prevent damage to the tree. The fixing can be checked every two or three years, to make sure the tree hasn't grown too broad for the wire. Tilt the box forward slightly so that any driving rain hits the roof and bounces clear.

If you put a nest box up in autumn, Wrens can use it to roost inside through the winter. You might even find a wood mouse taking shelter. Winter is a good time to put up a nest box too – fix it in place by the end of February, so that birds can become familiar with it before the breeding season begins.

In February and March, look out for pairs of birds investigating the box. The male enters first, then comes out to allow the female to look around. Watching a family of tits nesting in a garden box is a daily delight.

The spring equinox, around 21st March, is a point of perfect balance between winter and spring. Balancing can involve wobbling, and the month of March can swing back and forth between winter and spring, sometimes in the space of a single day! It can be wet and cold, grey as a winter's day; March winds can shake the daffodils and sweep the sky. But when the sun shines, dandelions and celandines open their yellow flowers, and suddenly it feels like spring.

The birds know spring is in the air – it is time for them to make nests. In this chapter we learn a little about nesting, and meet the bird that has always nested alongside humans, the House Sparrow.

MARCH

Nest Building

Early birds – especially older, more experienced birds of species like Robins, Blackbirds and Song Thrushes – often begin nesting in March. Look out for Blackbirds and Song Thrushes on lawns and in hedge bottoms, gathering dry grass and moss. If there are free-range hens in a garden or farm nearby, you might even spot Long-tailed Tits collecting beakfuls of feathers.

Depending on the species, the male or female bird may make the nest, or both may work together. Some birds make use of the old nests of other species. Opportunist House Sparrows sometimes take over a House Martin nest, moving in before the summer visitors return. In Europe, where Storks nest, a whole colony of little House Sparrows can nest in the base of the Stork's gigantic nest.

In March, it's easy to spot larger birds, like pigeons and corvids, carrying twigs for their nests. At this time of year, the rookery is loud with activity. The male bird brings a twig and is met with a noisy greeting.

Open-topped or cup-shaped nests need protection from the weather, so these nests are carefully placed in a nook where leaves and branches form a roof overhead, to keep growing chicks warm and dry.

Some birds, like Red Kites, even 'decorate' their nests with bright finds, including washing hung out to dry. Starlings line their nests with plants that deter lice. Blue Tits add herbs like mint or lavender – natural disinfectants.

We humans are gifted with clever hands and nimble fingers. It is truly a wonder that birds craft their baskets, balls and bowls using only their beaks.

Near roof-tops, look out for Jackdaws bringing sticks, grass or wool to their chimney-pot nests.

'THE BEST NEST' A FOLKTALE FROM ENGLAND

A story celebrating the nest

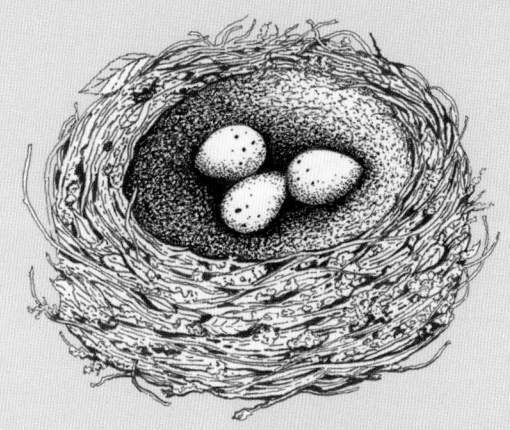

Song Thrush nest

The birds wanted to learn how to make a nest. A Magpie's nest is only equalled by a squirrel's drey, so it was Magpie they asked to teach them.

Magpie pulled a strong supple stick from the tip of a branch. 'Now', said Magpie, 'put one stick this way ... one stick that way ... one stick this way ... one stick that way ...'

Blackbird and Song Thrush watched carefully, nodding their heads. But Woodpigeon wasn't paying attention. He was busy scratching his neck with his foot.

'Now', said Magpie, 'you can add some mud to your nest ...' He flew to the edge of the pond. He scooped up a beakful of sticky mud. 'Smooth it this way ... smooth it that way ...'

Song Thrush nodded her head. She didn't wait to hear any more. Off she flew. And so, from that day to this day, that's the way Song Thrush makes her nest – lined with smooth bare earth.

'Now', said Magpie, 'you can add some soft grass and leaves ... one leaf here ... one leaf there ...'

Blackbird nodded her head. She didn't wait to hear any more. Off she flew. And so, from that day to this day, that's the way Blackbird makes her nest – lined with soft leaves and fine grass.

Magpies' nests and squirrels' dreys are both bigger than a football, so easy to spot in bare branches. Dreys are usually built close to the tree trunk, whereas Magpies' nests are often out on a branch. Dreys include twigs with leaves still attached, whereas the outside of a Magpie's nest includes only bare sticks.

Blackbird nest

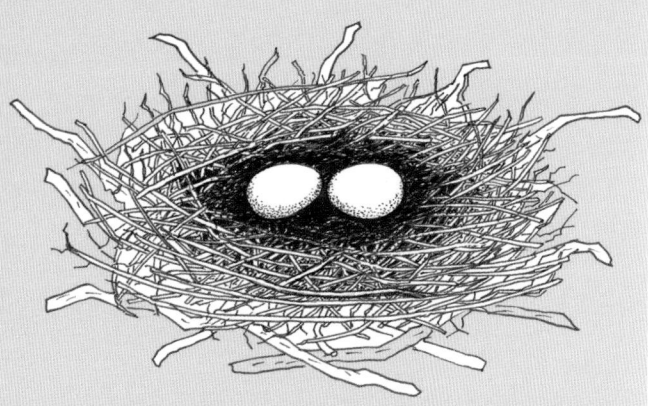

A Woodpigeon's nest – a platform of sticks – is stronger than it looks. They survive in treetops long after the family has moved on. Spot old Woodpigeon nests (and Carrion Crow nests – robust stick structures high up in tall trees) in early winter, before the wind and rain blow them away.

'Now, Pigeon,' said Magpie, 'I'll show you how to make a roof, to protect your babies. Put one stick this way … one stick that way … one stick this way … one stick that way …'

'What a bother!' thought Pigeon. 'There must be an easier way.' He interrupted, 'Take two-ooo sticks, Mag-gie!'

'One at a time's enough,' said Magpie. 'One stick this way … one stick…'

'Take two-ooo sticks, Mag-gie!'

'One's enough, I told you!' scolded Magpie. 'Put one stick …'

'Take two-ooo sticks, Mag-gie!'

'That's it!' cried Magpie. 'I've had enough!' And with a chattering cry, off he flew.

So Woodpigeon never did learn how to make a fine nest. From that day to this day, Woodpigeon's nest is nothing more than a slap-dash stack of sticks. But Woodpigeon doesn't mind. It may not be as smart as Magpie's nest, but it's just right for baby pigeons.

PLANT TREES, HEDGES AND SHRUBS FOR NESTING BIRDS

Birds rely on trees, bushes and hedges for safe shelter and for nest-sites. Trees and bushes also support insects and bear berries – providing birds with vital food. The total area of Britain's gardens is larger than the total area of Britain's nature reserves. So use your garden well, however big, or small, it is.

Choose native species, which offer most support to native wildlife. For example:

- Hawthorn, with its dense thorns, provides excellent protection from predators and is probably the favourite nest-site of hedgerow birds. Its May blossom supports insects and its abundant berries provide birds with an autumn feast. Hawthorn and hazel make a good base for a mixed native hedge.
- Ivy provides nest-sites for Sparrows, Robins and Wrens. Its flowers attract insects and feed caterpillars (providing food for nestlings) and its late-fruiting berries feed thrushes, Blackbirds and Woodpigeons through the bare months between January and March.
- Other native trees that support birdlife include: oak, ash, birch, beech, willow, hazel, blackthorn, crab-apple, rowan, cherry, elder, holly and yew.
- Other native shrubs that support birdlife include: bramble, dogwood, honeysuckle and wild privet.

MAKE A NEST-MATERIAL BALL

You can put nesting materials out for the birds in a fat-ball feeder, a hanging basket, or an old teapot hung up by the handle. Or you can make your own nest-material ball:

You will need:

- Bendable wire (without sharp edges), e.g. craft wire, gardening wire, even an old wire clothes hanger
- Wire cutters
- A ball
- Nesting material. Different birds use different materials, so include a variety, such as:

  - Wool (unspun, not thread)
  - Twigs and sticks
  - Dead leaves
  - Old grass clippings and dead grass (avoid any material that has been treated with chemicals)
  - Moss and lichen
  - Straw and plant stems (again, avoid any material that has been treated with chemicals)

1 Cut a long length of wire.
2 Wrap the wire around the ball. Imagine the ball is an orange, with the stalk at the top. Weave the wire once around the vertical circumference of the ball, twisting together at the top to secure.
3 Weave around the circumference again and again (as if making orange segments).
4 Bend the wire out of shape to remove the ball. Re-form your shape.
5 Add a hook for hanging.
6 Fill your ball with nesting materials.
7 To keep the nesting material dry, hang your ball somewhere that's sheltered from the rain. Ideally, position the nesting material in a different area from your bird feeders.

Safety note: If birds or other wildlife look as if they might get tangled in any feeder or nest-material holder, take it down to avoid possible injury.

*Note: To avoid disturbing nesting birds and developing fruits, do not trim hedges between March and August – ideally cut back during January and February.

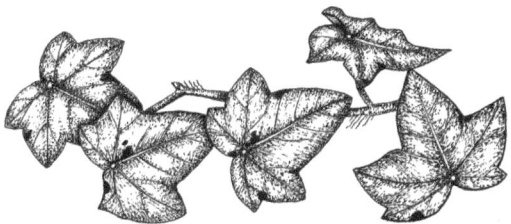

HOUSE SPARROW

Passer domesticus

Grey crown

Black 'bib'
– like a beard

Streaky brown
wings

MALE

Cream stripe behind eyes
– like eyeshadow

Darker stripe below
– like eyeliner

Streaky
brown wings

FEMALE

In flight, the male's broad
white wing-bar is clear.
Juveniles look similar to females.

VOICE
Speaks with a one-note *chirp*
that is used as a contact call between
mates or with others in the flock. In
the breeding season, the male strings
together his chirps and cheeps to
form a simple song, repeating it
loudly from rooftop or gutter.

SIZE
14cm

WHERE
Near buildings,
alongside
humans.

NEST
Hole in building.
Pairs often use
the same nest
site for life.

EATS
Seeds,
whatever's
available.

HOUSE SPARROW – THE OLD NAMES

Sparrow – from Anglo-Saxon *spearwa*, stemming from Icelandic *sporr*, 'flutterer'.
Variants include: **Sparrer** (as in 'Cockney Sparrer') • **Spadger** (Lancashire, Suffolk)
Spadge – sometimes used as a term of endearment in Derbyshire • **Spurdie** (Orkney Isles)

Sparrows nest under house eaves, hence the names:
Thatch Sparrow (Northamptonshire, Shropshire) • **Eaves Sparrow** (Shropshire, Nottinghamshire)
Hoosie (Northumberland) • *Aderyn y To* – 'roof bird' (Wales)

From the Sparrow's *phip* chirrup: **Philip**

SPARROW OF THE HOUSE

House Sparrows are well named – no bird has lived alongside humans for so long. We've been neighbours for at least 10,000 years, since Stone-Age times. When Neolithic hunter-gatherers first began to build homes and farm the land, seed-eating House Sparrows ate the spilt grain on their farms, and nested in the thatch roofs and walls of their buildings. It is thought that, in Britain, there were no House Sparrows before humans arrived. When Neolithic peoples spread across Europe from their original home in Africa, House Sparrows came with them. They've lived alongside us ever since.

When most farmhouses and cottages were thatched, the spaces under the eaves were a House Sparrow's favourite nesting place. The male finds the nest-site, and calls beside it to attract a mate. When a female comes close, he flutters his wings and lifts his head – showing off his black 'bib'. The female lines the nest-hole, using whatever she can find, even scraps of paper and feathers plucked from live pigeons.

House Sparrows often nest in groups, and pairs often remain faithful to the same nest-site, as well as to each other. A spattering of droppings high up on your brickwork, underneath the roof tiles, can give you a clue that your home is also home to some feathered friends.

House Sparrows have lived alongside humans in all kinds of unlikely environments. Sparrows have been found in the depths of coal mines and at the heights of the 80th floor of the Empire State Building. In the Second World War, House Sparrows followed army camps across the North African desert. Sparrows famously nested in the warehouse of Penguin Books Ltd, alongside the bookshelves full of puffins and pelicans. They've even been known to take rides on London Underground trains and cross-Channel ferries, and to open automatic doors to get into supermarkets. In town centres, flocks of Sparrows often feed on leftover crumbs. They are so adapted to living alongside humans that where they're fed regularly they can become tame and even be fed by hand.

HOME-LOVING SPARROW

For much of the year, it's rare to find House Sparrows far from human homes. Few House Sparrows ever journey more than a mile or so from where they were hatched and raised, staying in the same

room patch for most of the year. The only time they venture further is in August. While we're heading off for a summer holiday, House Sparrows may also leave town to travel further afield; literally – to a nearby field, to feed on ripening grain. They feast for a week or two, before, like us, returning home again. Juvenile birds, born in the spring, often make up a large number of these late-summer flocks. They are looking for homes of their own, somewhere less than a mile or so from where they were born, before, like their parents, they settle down to stay in the same home for the rest of their lives.

A TRIBE OF SPARROWS

In urban streets, you might hear a bush chirping noisily, as House Sparrows call to each other within. House Sparrows are almost always seen (or heard) in groups. As well as nesting in a colony, from late summer and through the winter House Sparrows also roost in groups, in shrubs, hedgerows or buildings. In colder areas, they may even roost on streetlamps for extra warmth.

DRY CLEANING – THE DUST BATH

As well as enjoying a splash in the bird bath on a summer's day, you can often see a Sparrow party enjoying a 'dust bath'.

In high summer, when it's very hot and dry, the birds loosen a patch of crumbly earth with their claws, then squat in the 'bath' and ruffle their plumage, shaking their feathers and flicking their wings to send up clouds of dust. They work the fine dry earth into their feathers, just as we might dust our skin with talcum powder. The Sparrows finish by shaking out the dust and preening their feathers.

Clearly dust bathing is part of the birds' feather-care routine, though it's not fully understood. One theory is that in hot weather birds produce more preen oil (just as humans might sweat more). It is thought that the dust works like a 'dry shampoo' to remove some of the oil from the birds' feathers. Dust bathing is also thought to remove lice and mites.

Give the House Sparrows a spa day by making a dust bath for them. Choose a sunny spot with shelter nearby and create a hollow in the earth (about 60cm by 30cm – big enough for a whole party of birds to enjoy). You could contain the hollow with a decorative ring of rocks. Fill it a few centimetres deep with sieved earth, sand and wood ash. Making a dust bath is a simple job, perfect for very young helpers, who will enjoy the pleasure of getting their hands in the earth, digging, sieving, patting and smoothing.

YELLOW FLOWERS

In February and March, when the first brave primroses and crocuses add a splash of sunshine to our lawns and verges with their yellow flowers, you might spot a House Sparrow tearing at the petals with its beak. It has long been observed that House Sparrows peck and tear at yellow spring flowers, but the reason for this remains a mystery. One thought is that yellow flowers are attractive to Sparrows because they contain the yellow-orange pigment carotene, to brighten up the bird's plumage for the breeding season. Another idea is that the yellow colour of so many spring flowers is a bright sign to attract pollinating insects, since yellow flowers tend to be rich in nectar (yellow crocuses, for example, contain more nectar than purple ones). Perhaps it's this sweet nectar at the base of the flower that Sparrows are enjoying when they peck your primroses. Primroses are edible flowers for humans too. In spring, children can enjoy snacking on yellow primrose petals, just like their friends the Sparrows.

LOVERS' BIRD

The male House Sparrow's black bib is his mark of masculinity. In winter, the male's bib (and his grey crown) are paler, flecked with white. By spring, the tips of the bird's feathers have worn away, revealing his new breeding plumage beneath. With perfect timing for courting, he sports a smart grey cap and a bold, black 'bib' – like a magnificent beard! The bigger the male's bib, the more attractive he is as a mate. Studies even showed that when males with small bibs had bigger, blacker bibs painted onto them, they behaved more boldly.

In the spring, look out for the male's courtship display. The male bird droops his wings and cocks his tail, and puffs out his handsome bib, showing off to the female and bowing to her (though if he gets too close, she might peck him). Sometimes other males nearby join in, all chirping loudly in the hope of attracting the female's attention.

It's no wonder that male House Sparrows are keen to mate – during the breeding season the testes of the male birds enlarge by between 300 and 500 times (outside of the breeding season, they weigh less to aid flight).

Most pairs of House Sparrows will raise at least two, often three, broods a year. Because they mate out in the open, in ancient Egypt the Sparrow was the hieroglyph for 'lustful', and in ancient Greece they were associated with Aphrodite, goddess of love. Aristotle said 'of all the birds [they are] the most wanton'. Despite Aristotle's disapproval, House Sparrows' eggs were a popular aphrodisiac. In 1559, in Saxony, House Sparrows were even banned from church, for their 'endless unchaste behaviour' before the altar of the Holy Cross Church in Dresden.

Because the birds usually mate for life, they were also an omen of marriage. If an unmarried woman sees a Sparrow on 14th February, she will wed a poor man within the year, yet will find happiness.

'THE SINGING SPARROW' A YIDDISH FOLKTALE

A story that highlights the repetitive quality of the House Sparrow's simple song

Up on the rooftop, Sparrow's cheerful chirping was loud and long – it went on and on and on! Crow was fed up with it. 'Why are you chirp-chirp-chirping so much?' he snapped. 'What are you so cheerful about, anyway?'

'I have plenty of seed in my little territory,' said Sparrow. 'Cheep, cheep, cheep!'

Crow frowned. 'It could stick in your throat and choke you!' he said, keen for some quiet.

'I'll just scratch it out. Chirp-cheep!'

'And make your throat bleed!' said Crow, with relish.

'I'll drink water to make it better. Chirp-chirp-chirp-chirp, cheep!'

'Cold water!' cawed Crow. 'You'll catch your death of cold!'

'I'll make a fire to warm up. Chirp-cheep, chirp-chirp-cheep!'

'But the fire could spread!'

'I'll just put it out with my wings. Chirp, cheep-cheep!'

'But your wings could get burnt!'

'I'll go to the doctor, then. Chirp-chirp, cheep-cheep!'

'Well,' said the Crow, in exasperation, 'what if there is no doctor?'

'Then I'll get well, without a doctor!' said the Sparrow, cheerful and chirpy as ever. 'Chirp-cheep, chirp-cheep, chirp-cheep, chirp-cheep!'

Crow groaned. He flapped away – there's no changing Sparrow's song!

HELP HOUSE SPARROWS THRIVE

According to the RSPB's annual survey, the Big Garden Birdwatch, House Sparrows have been the most numerous bird spotted in UK gardens for the past 20 plus years. But the survey also showed that between 1977 and 2008 the UK House Sparrow population dropped by 71 per cent.

Various reasons have been suggested, such as loss of suitable nest-sites due to changes in building styles. Old houses had spaces under the thatch-eaves or tiled roofs, which recent building designs have lacked. Another cause seems to be lack of insects to feed nestlings. Intensive farming methods (such as monoculture fields and pesticide use) and the loss of native plants in cities, and in gardens (to make way for car-parking spaces, for example), have all impacted insect populations.

In towns and cities, other causes suggested include electromagnetic radiation from mobile phones and air pollution from traffic. Like the miner's Canary, House Sparrows give us a clear warning sign about the harmful effects of many modern norms on human health, as well as on the health of all life.

In 2020, the results of the Big Garden Bird-watch kindled a spark of hope for the House Sparrow. In Scotland, Wales and Northern Ireland, garden sightings of the bird were up by 2 per cent since 2019. Let's hope this marks a turning point for our friendly feathered neighbour.

Our actions have the power to help fuel that shift. Our food-buying choices can be investments into wildlife-friendly farming. Our homes can offer places for House Sparrows to nest. To adapt the bird box on pages 48–49 to suit Sparrows, simply cut the entrance hole to a size of 32mm. As House Sparrows often nest in a loose colony, you could put up two or three boxes on the same patch of wall, to create a 'Sparrow street'.

Our gardens can be feasting grounds for hungry birds. Planting native shrubs and trees (page 54) provides Sparrows with vital cover, insects and berries. Bird-friendly flowers you could plant include:

- Early-flowering bulbs, like snowdrops, to attract insects (yellow flowers, like crocuses and primroses, are especially attractive to House Sparrows).
- House Sparrows are seed-eaters, so seed-bearing flowers (like sunflowers) provide good food. Thistles and teasels provide seeds and insects (which winter in the stems).
- Weeds also provide seeds that Sparrows eat. Chickweed is a favourite. Embrace the wild and leave the weeds to grow (or at least *some* weeds).
- Perennials (especially if left uncut over winter).

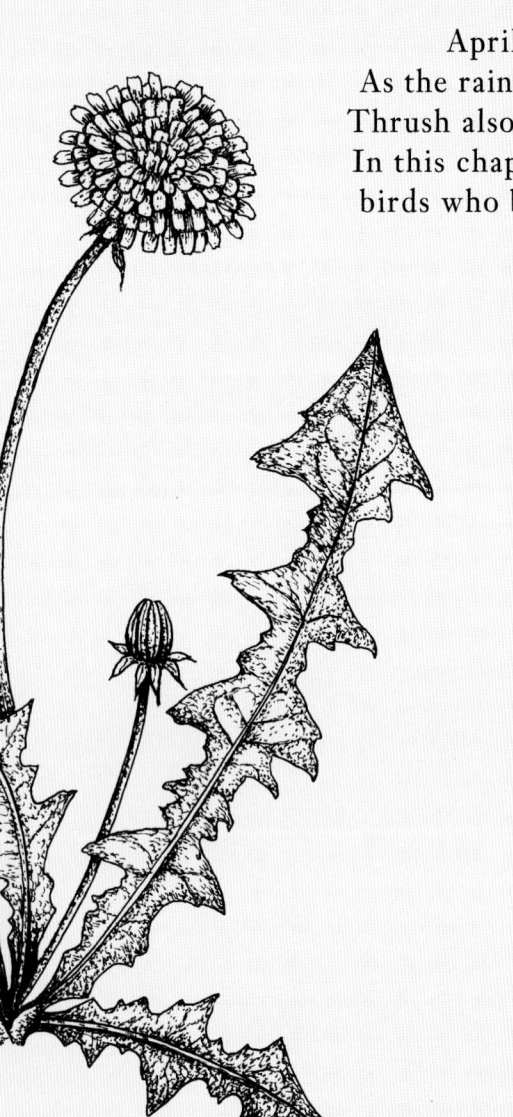

April. Sunshine and showers. And flowers, in
colours as sweet as Easter eggs. Blossom, fat pink
pom-poms and delicate rose-tinted apple petals. Bluebells,
deep pools of blue in the green light of the woods.
A treasure-hoard of buttercups, shining in the sun.
And the sound of birdsong.

April showers don't deter the Blackbird.
As the rain falls, the Blackbird's song rises. The Song
Thrush also appreciates the rain that serves the snails.
In this chapter we celebrate these close cousins – two
birds who brighten our days with their spring songs,
whatever the weather.

APRIL

BLACKBIRD

Turdus merula

Yellow eye-ring

Bright yellow-orange
beak

MALE

Glossy black
feathers

Dark
beak

Brown feathers

Speckled
breast

FEMALE

See page 92 for
picture of juvenile.

VOICE

Sings from a high perch.
Liquid song is clear and fluting,
melodious and mellow paced. Notice
the short pauses between phrases.
Alarm call is a scolding *chip chip chip*.
When agitated, the call speeds to
a loud, rattling *chipchipchip*, often
as the bird explodes from cover
and takes flight.

SIZE

25cm

WHERE

Gardens, parks,
hedgerows,
woodlands, open
lands.

NEST

Grass-lined cup
of moss and
mud, low down,
sheltered by
leaves. Female
incubates chicks,
both parents
feed them.

EATS

In summer,
eats insects
and worms.
In winter, eats
berries.

BLACKBIRD – THE OLD NAMES

DARK BLACK AND BRIGHT GOLD

Many thrushes have speckled breasts, but the male Blackbird, as the name suggests, is glossy black all over. According to an Italian folktale, the bird was once white, but the last two days of January, and the first day of February, were so cold the bird took refuge in a chimney. He's had sooty black feathers ever since, and those three freezing days are known as '*i giorni della merla*' – the Blackbird's days.

A French tale says that the Magpie told the Blackbird where to find treasure, deep in a cavern. The Blackbird turned over the flakes of gold with his beak, leaving it golden-bright. (Blackbirds do turn over fallen leaves, when they're looking for food – listen out for the rustle of a Blackbird foraging.) The story adds that the bird was disturbed by a treasure-guarding monster, breathing fire and smoke, which blackened his feathers forever. The Blackbird took flight with a loud cry – the same loud cry he still makes today.

In Welsh mythology, the Blackbird of Cilgrwi proves his great age by explaining that the metal stump he sits upon was once a Blacksmith's anvil. Over the ages, it was worn down by the rubbing of the Blackbird's beak. You may spot a bird cleaning its beak the same way – by rubbing it against a branch. Perhaps the connection between Blackbirds and blacksmiths stems from the bird's black feathers and bright beak and the black soot and bright fire of the blacksmith's craft. Female birds are more attracted to a male with a bright beak. The colour comes from carotenoids in food – a bright beak shows that the bird is good at foraging, which will benefit the whole family.

SINGING IN THE RAIN

Rain brings worms to the surface of the earth (as it's easier to wiggle over the ground when it's wet). Blackbirds love earthworms – they're a baby Blackbird's favourite food.

Blackbirds can hear movement under the ground – notice how they cock their heads as they listen and look for worms.

The birds also need rain for the mud used in their nests. Blackbirds can often be heard singing after rain. Their large eyes help them see in the half-light of dusk, so you can hear them singing as the sun sets too. Blackbirds are superb songsters – they've even been known to incorporate tunes whistled by people into their songs.

Not only do Blackbirds sing after the rain, they also do rain-dances. See if you can spot a Blackbird 'paddling' the ground with its feet – the footsteps send a vibration down into the earth, like the vibrations made by pattering raindrops, to draw worms up to the surface.

Blackbirds often hop across the ground on both feet at once, leaving paired tracks, side by side.

BASKING IN THE SUN

You might also spot a Blackbird sunbathing. In hot weather, the Blackbird sits on the grass or warm ground, basking in the sun with feathers fluffed and wings fanned. The heat of the sun helps the bird's 'preen oil' spread along its feathers, keeping them strong, supple and waterproof. The heat also drives out any lice and parasites from the bird's feathers, so the bird can remove them with its beak.

BIRDS AND BERRIES

Blackbirds love berries. In summer, strawberries and cherries are particular favourites.

I value my garden more for being full of blackbirds than of cherries, and very frankly give them fruit for their songs.

JOSEPH ADDISON (FOUNDER OF *THE SPECTATOR*)

After the summer, when there are fewer insects, Blackbirds feed mainly on fruit. In winter months, the late berries of ivy are especially appreciated. Look out for a Blackbird perching in a bush or climber, as it reaches for ripe berries.

Birds and berries enjoy a special friendship. The reason berries are so bright and sweet is to attract birds (and other creatures) to eat them. The plant gives the bird delicious, nutritious food. And in return, the bird carries the seed of the plant in its belly to a new home. It even gives the seed a helping hand to grow, by dropping it with a healthy dollop of fertiliser, otherwise known as bird poo!

BLACKBIRD NEST

A Blackbird nesting near a house is a sign of good fortune. An Irish myth tells that St Kevin was once praying with arms outstretched. A Blackbird landed in his open hands and there laid her eggs. The compassionate saint kept his arms outstretched until the eggs had hatched and the brood flown.

BAKED IN A PIE

In past times, when food was scarce, Blackbirds were sometimes hunted for meat. In winter, Blackbirds spend the night in communal roosts, sheltered in the thick undergrowth, and native birds are joined by migrants escaping the freezing winters of northern Europe. (On mild evenings, you might hear the *pink, pink, pink* of Blackbirds settling down to roost.) With so many Blackbirds gathered together, a night-roost was an easy place to trap birds. In medieval times, live birds were placed just under the pie crust before serving:

Sing a song of sixpence,
A pocket full of rye;
Four and twenty blackbirds baked in a pie!
When the pie was opened the birds began to sing,
Wasn't that a dainty dish to set before the king?

'BLACKBIRD AND HIS WIFE'
A FOLKTALE FROM INDIA

A story in praise of the Blackbird's song – the tale also introduces the idea that male and female birds can have different feathers

Once upon a time there was a Blackbird. He loved to sit at the very tip-top of a tall tree, and sing. The Blackbird's song was clear and bright as water from a spring.

Well, one day the king heard the Blackbird singing and he wanted that song for himself. So he sent his men to catch the bird. But the king's men didn't know the difference between the Blackbird and his wife, and they caught the brown bird – the female. The king kept her in a golden cage, but she never sang a note.

And when the Blackbird saw that, he was *so* cross. The Blackbird found a nut. *Crack!* From one half he made a helmet. From the other half he made a drum. He beat the drum. And he set off, marching to the palace of the king. *Rat-tat-tat! Rat-tat-tat!*

On the way, he met a fox. A sleek red fox with dark ears and a white tail tip.

'Where are you going, beating your drum?'

'The king has taken my wife. I'm going to set her free.'

'Well, I'll come with you,' said the fox. 'For years the king has hunted me and hounded me. I'll help you.'

'Then jump into my ear', said the Blackbird, 'and I'll give you a ride.' So the fox shrank itself, teeny-tiny, and leapt into the Blackbird's ear. And off he went on his way. *Rat-tat-tat! Rat-tat-tat!*

He hadn't gone far before, there on the ground, he found a rope. A bit of old rope, faded and frayed.

'Where are you going, beating your drum?'

'The king has taken my wife. I'm going to set her free.'

'Well, I'll come with you,' said the rope. 'For years the king has twisted me and tangled me. I'll help you.'

'Then jump into my ear,' said the Blackbird. 'I'll give you a ride.' And off he went. *Rat-tat-tat! Rat-tat-tat!*

He hadn't gone far before he came across a great heap of red earth – an ant-hill. And a trail of ants, walking in a line, one behind the other.

'Where are you going, beating your drum?'

'The king has taken my wife. I'm going to set her free.'

'Well, we'll come with you,' said the ants. 'For years the king has stepped on us and stamped on us. We'll help you.'

'Then jump into my ear,' said the Blackbird. 'I'll give you a ride.' *Rat-tat-tat! Rat-tat-tat!*

He hadn't gone far before he met a river.

'Where are you going, beating your drum?'

'The king has taken my wife. I'm going to set her free.'

'Well,' said the river, 'I'll come with you. For years the king has dammed me and dirtied me. I'll help you.'

'Then jump into my ear,' said the Blackbird. 'I'll give you a ride.' *Rat-tat-tat! Rat-tat-tat!*

Before long, he came to the palace of the king. *Rat-tat-tat! Rat-tat-tat!* 'I've come to free my wife!'

But the king just laughed. 'You? Ha!' He turned to his guards. 'Throw him in with the chickens.

They'll peck him and scratch him and that'll be the end of that.'

But in the hen house, the Blackbird gave a whistle. And out leapt the fox. He growled and he snapped. Squawking and flapping and feathers flying! And in the morning, when the king opened the door, he found the chickens cowering in the corner, and the Blackbird beating his drum. *Rat-tat-tat! Rat-tat-tat!*

'What?!' The king was furious. 'Throw him into the stable, with the horses,' shouted the king. 'They'll stamp and they'll kick, and that'll be the end of that.'

But in the stable, the Blackbird gave a whistle. And out came the rope. It curled and coiled, and cracked like a whip. And in the morning, when the king opened the door, he found the horses cowering in the corner, and the Blackbird beating his drum. *Rat-tat-tat! Rat-tat-tat!*

'What?!' The king was fuming. 'Throw him in with the elephants! They'll trumpet and trample, and that'll be the end of that.'

But in the elephants' pen, the Blackbird gave a whistle. And out came the ants, teeming and seething. They crawled right inside the elephants' trunks, and tickled with their feet and prickled with their stings. And in the morning, when the king opened the door, he found the elephants cowering in the corner, and the Blackbird beating his drum. *Rat-tat-tat! Rat-tat-tat!*

'What?!' cried the king. 'What magic is this? Tie this bothersome bird to my bedpost, and I'll find out!'

That night, the king lay in bed, listening. The Blackbird gave a whistle. And out rushed the river. It swelled and swirled and filled the king's room with water. The water washed over the king's carpet. The water rose higher and higher, filling the room. Soon the water was washing over the king's bed – the bed began to float. The king's pyjamas were sopping wet. The king was soggy and shiv-

ering. His bed was rocking and swaying. His crown was washed clean out the window.

Rat-tat-tat! Rat-tat-tat! The Blackbird beat his drum. 'Let my wife go free!'

'Take her!' said the king, 'and go!' The king cut the knot from the bedpost and he opened the door of the golden cage, and the Blackbird and his wife flew free. They flew over the lawns and over the gardens, over the fields and over the hedges, back to their own tree. The Blackbird's wife hopped around the ground, finding good things to eat. And the Blackbird sat at the top of the tree – he opened his beak and he sang. And the Blackbird's song was clear and bright as water from a spring.

SONG THRUSH

Turdus philomelos

Speckles
shaped like upside-
down hearts
– or arrowheads,
pointing upwards

Warm brown

White belly

Juveniles have backs that are
more speckled than adults'.

Smaller than
Blackbird

VOICE
Sings from a high perch, in a
loud clear voice, repeating the same
series of notes (or 'phrase') over and
over. A Song Thrush's alarm call
is similar to a Blackbird's – a loud
chuck-chuck. Often gives a short,
high-pitched *tsip* in flight.

SIZE
23cm

WHERE
Gardens, parks,
hedgerows,
woodlands.

NEST
Mud-lined
cup-nest, low
down, sheltered
by leaves. Female
incubates chicks,
both parents
feed them.

EATS
Worms, snails,
bugs, fruit,
berries.

APRIL

SONG THRUSH – THE OLD NAMES

Throstle – from Anglo-Saxon (Northern England, Midlands, Ireland)
Thrushel/Thrustle (Shropshire) • **Thrush Drush** (Somerset)

Thrice Cock – the Song Thrush often repeats the same note three times (Old English)

Mavis – from French *Mauvis* (East Anglia, Ireland, Scotland)

THE POET'S BIRD

The Song Thrush lives up to its name, with a loud, long song – lasting five minutes or more. A Song Thrush's repertoire contains around a hundred different 'phrases'. Some of these are learnt from neighbouring birds. Some are learnt from previous generations. It is wonderful to think that the Song Thrush phrases we hear today have been heard by many generations before us. Rather than singing different phrases one after the other, a Song Thrush repeats the same phrase several times, then switches to a different phrase, and repeats that over and over again.

The music of the thrush has inspired poets through the ages. Thomas Hardy described it as 'full-hearted evensong'.

'The Speech of Birds', in Alexander Carmichael's collection of Gaelic hymns and incantations, *Carmina Gadelica*, includes a verse evoking the Song Thrush's repetitive song:

The Mavis Said:
Little red lad!
Little red lad!
Come away home!
Come away home!
Come away home,
My dear, to your dinner!

What shall I get?
What shall I get?

A worm and a scrap of limpet!
A worm and a scrap of limpet!

Hurry up! Hurry up!
The night's coming!
The night's coming!
And the darkness!

William MacGillivray also echoes the Thrush's sing-it-again style:

Dear, dear, dear
Is the rocky glen;
Far away, far away, far away,
The haunts of men.

FROM 'THE THRUSH'S SONG'

As does Tennyson:

'Summer is coming, summer is coming.
I know it, I know it, I know it.
Light again, leaf again, life again, love again,'
Yes, my wild little Poet.

FROM 'THE THROSTLE'

Robert Browning says:

That's the wise thrush; he sings each
song twice over,
Lest you should think he could never recapture
The first fine careless rapture!

And William Wordsworth shares these words
of wisdom:

Hark, how blithe the throstle sings
And he is no mean preacher
Come forth into the light of things
Let Nature be your teacher.

SNAIL STONE

Like its cousin the Blackbird, the Song Thrush
enjoys earthworms, running along the ground and
stopping to listen and look for worms and bugs. It is
especially fond of snails. When hard ground makes
worms less easy to find, you might spot a Thrush's
anvil – a large stone surrounded by broken snail
shells. Of all garden birds, only Thrushes can open
a snail shell, and each Song Thrush has its own
snail-stone. The bird holds the snail in its beak,
and with a quick flick of its head, smashes the shell
against the stone to get to the juicy food inside.
Listen out for the tapping sound of a Song Thrush
using its anvil.

AUTUMN ARRIVALS

Song Thrushes often give a short, high-pitched *tsip*
in flight. British Song Thrushes tend to stay close
to home, but in late autumn, birds from northern
Europe do migrate to escape the cold. Some spend
the winter in Britain and Ireland, and others pass
through, as they journey further south. The birds
fly by night in loose flocks, calling to each other
to keep in contact – on October evenings, listen
for the calls of night-flyers, as they wing their way
overhead.

Irish folklore tells us that the Thrush's song is
loved by faeries, as well as poets. It was said that
the little folk encouraged 'Mavis' to build her nest
low down, near to the ground, so that they could
hear and enjoy the bird's music.

'SINGING IT OVER AND OVER' A FOLKTALE FROM INDIA

A story to tune our ears to the repetitive rhythm of Song Thrush's song

Once upon a time there was a Song Thrush who loved to sing. He loved singing so much, he always sang the same little ditties over and over again:

> *Di-dit, di-dit, di-dit,*
> *Wheet, wheet, wheet,*
> *De-dew, de-dew, de-dew ...*

Well, one day, the king was out hunting, hoping to catch a bird or a beast for the pot. And when he heard the Thrush's loud, clear song, he took aim. *Whew!* His arrow flew. *Thwack!* He hit the bird. *Thud!* Down it fell. And the king scooped it into his sack, and carried it back to the castle.

But, as he walked, and the sack thumped against his back, from inside the sack came the sound of singing:

> *Di-dit, di-dit, di-dit,*
> *Wheet, wheet, wheet,*
> *De-dew, de-dew, de-dew ...*

Well, the king was surprised to hear the bird still singing, but even so, he took the bird to the cook. And the cook began to pluck the bird. But every feather the cook plucked flew – *whoosh!* – out of the open window, all by itself! And when the cook put the bird in the pot, from inside the pot came the sound of singing:

> *Di-dit, di-dit, di-dit,*
> *Wheet, wheet, wheet,*
> *De-dew, de-dew, de-dew ...*

When the bird was done, the cook put it on a plate and set the plate in the middle of the table. But as soon as the king picked up the carving knife, the bird began to sing again!

> *Di-dit, di-dit, di-dit,*
> *Wheet, wheet, wheet,*
> *De-dew, de-dew, de-dew ...*

At that very moment there was a great chorus of calls, and there at the window was a whole flock of birds. And in each beak there was a feather. Warm brown feathers and soft cream feathers, and speckly, freckly feathers patterned with little hearts. And – *whoosh!* – with a rush of air, the feathers flew into the room and the Song Thrush shook out its wings – whole again. Up he rose and off he flew, out of the window. He sat at the top of a tall tree and he sang – a loud and clear song, singing the same sounds, over and over again:

> *Di-dit, di-dit, di-dit,*
> *Wheet, wheet, wheet,*
> *De-dew, de-dew, de-dew ...*

SONG THRUSH

HELP SONG THRUSHES THRIVE

Between 1970 and 1998, Song Thrush populations declined by 59 per cent. Causes include loss of habitats, like cow pastures and woodlands, and the use of pesticides that poison the bugs the Song Thrush eats.

At the time of writing, the Song Thrush is on the 'red list' of Birds of Conservation Concern. During the last decade there have been hopeful signs of a partial recovery in numbers, and there are things we can all do to fuel that hope and help the Song Thrush thrive.

Poisonous chemicals – pesticides, insecticides, herbicides – harm wildlife. In the midst of lost and degraded habitats, our gardens (and our verges too) can offer a vital oasis. Using poisons like slug pellets in our gardens reduces food supplies for Song Thrushes, and other wildlife. Instead, we can respond to slugs and snails in bird-friendly ways:

- Remove slugs by hand. At night, go out with a torch and collect slugs and snails in a bucket. Put them somewhere where they can live without crossing paths with gardeners.
- Encourage creatures that feed on slugs. Support frogs with a pond and hedgehogs with 'hedgehog highways' of holes cut in fences. A log pile supports slug-eating beetles.
- Soft-bellied slugs and snails find it hard to move over certain surfaces. Try spreading (dry) crushed eggshells, wood ash, coffee grounds or wool pellets around plants. Plants in pots can be protected with a band of copper tape around the rim.
- Try companion planting – protect favourite garden plants with plants that slugs don't like, such as garlic or chives.
- Offer slugs an alternative. Keep slugs from eating your favourite garden flowers or vegetables by offering them a more attractive option instead. Lawn chamomile and lettuce are particularly alluring to slugs, keeping them, and you, happy.

MAKE A MINI POND

All birds need water, to drink and to bathe in (to keep feathers in good condition for flying). Even a dish of water on a bird table helps, but if you have outdoor space, a pond is ideal.

You will need:

- An old container (large, waterproof and weatherproof) e.g. washing-up bowl, garden trug, plant pot (lined with butyl pond-liner) or kitchen sink (an adult can silicone in a plug to seal the plughole)
- Small stones or gravel
- Pebbles, rocks or twigs
- Native water plants

1 Find a good spot for the pond. Position your container before filling with water, whilst it's not too heavy. Ideally, position it somewhere that gets some sun, but isn't in full sun all day. Positioning the pond amidst plants, rocks and pots provides cover for wildlife too. If you can, dig a hole to hold the container, so that the top edges are level with the ground – this helps creatures get in and out more easily (if you want a more decorative effect, you could surround the edges with attractive stones). You can also leave the container on top of the ground.

2 Prepare the container. Put a layer of clean gravel in the bottom (avoid soil, which makes the water go green). Make sure creatures can get in and out of the water by creating stepping stones and ramps, using stones, tiles, logs or sticks.

3 Fill your pond with water. Use rainwater – the chemicals in tap water are not good for wildlife.

4 Add water plants. You can get special aquatic plant pots (with mesh sides). Use a very low nutrient soil (you can also get special soil for ponds) mixed with grit. Always use native plants, and choose plants that won't grow too large for the space. Include a mix of submerged pondweed, to help the water stay clear (the RSPB recommends rigid hornwort and whorled water milfoil) and upright plants (like marsh marigold, lesser spearwort and water forget-me-not) positioned around the edge to provide perches and cover for wildlife. No more than two or three upright plants are needed for a mini pond.

5 Don't worry if, to start with, green gloopy algae or weed forms on the pond – you can scoop it out by winding it round a stick, a job children will enjoy.

6 If you need to top the pond up in hot weather, use rainwater.

7 Over time, the pond will develop an ecosystem of its own, and can help frogs and toads, hedgehogs and bats, damselflies and dragonflies, as well as birds.

May – the flower garland of the year. The world is singing green. The land is alight with white. Foaming sprays of Queen Anne's lace, and waving swathes of May. Stitchwort, embroidering country lanes, and daisies, cheering the grass. Even the dandelions have turned to white. The air is soft and scented, dancing with insects, ringing with birdsong. We rejoice in the dawn chorus.

The soothing crooning of doves is the soundtrack of summer afternoons – the drowsy sound of comfort and contentment. In this chapter, we meet three of our most familiar pigeons, the Woodpigeon, the Collared Dove and the Feral Pigeon.

MAY

WOODPIGEON

Columba palumbus

In flight, white wing-stripes are clear

White neck patch

Plum-pink breast

Plump body

Juveniles do not have a white neck patch.

VOICE
The soothing crooning of the Woodpigeon is easy to recognise – it has a throaty five-note song, with the emphasis on the second note: *coo coooo coo, coo-coo*. When disturbed, the clatter of wings as the bird takes off is unmistakable. The noise startles predators and creates a loud, clear warning.

SIZE
40–42cm

WHERE
Farmland, woods, parks, suburban gardens.

NEST
Platform of sticks in tree. Both birds make the nest. Male incubates the nest by day; female by night.

EATS
Seeds, leaves, buds, crops, berries.

WOODPIGEON – THE OLD NAMES

From the Woodpigeon's gentle coo:
Too-zoo (Gloucestershire) • **Doo** (Suffolk) • **Cushie Doo** (Yorkshire)
Coo Shot – literally 'coo-shouter' (from Anglo-Saxon *cūscote*) (Yorkshire)
Cushat (Scotland, Berkshire, Buckinghamshire)

From the Woodpigeon's noisy take-off: **Clatter Dove** (Yorkshire)

From the white neck-ring: **Ring Dove**

A PORTLY GENT

The Woodpigeon is the largest of the dove family – it walks with a sedate waddle, like a portly old gentleman. Woodpigeons look as if they enjoy their food, because they do! Originally a woodland bird (as the name shows), the Woodpigeon's appetite for farmed grain meant its numbers increased – it is now a common sight in parks and gardens. The phlegmatic Woodpigeon also enjoys a rest (and sometimes uses an old nest as a sitting-room). Notice how the bird fluffs up its plumage and draws in its head – snuggling down into a warm feathery scarf.

A FEAST ON THE FARM

In the woods, Woodpigeons love acorns, eating them straight from the tree whilst they're still green. Look for them in autumn oaks. A single Woodpigeon can eat 120 acorns in one day. They also enjoy beechmast, ivy berries and weed seeds.

Woodpigeons also enjoy the same foods that we do – cereal grains like wheat, rye and oats, and vegetables like potatoes, beans, peas, cauliflowers, radishes and greens.

In spring Woodpigeons eat newly sown crops, in summer they like grain, in autumn they feed amongst the stubble and in winter they feast in huge flocks on fields of leafy greens.

Woodpigeons clearly prefer the large grains of crops to the tiny seeds of wild grasses, known as 'bents'. When Woodpigeons are reduced to eating bents, it has been a bad year for farmers, as the old Norfolk proverb says:

> *When the pigeons go a benting,*
> *Then the farmers lie lamenting.*

SPECIAL SKILLS

Woodpigeons' appetite for hard, dry grains means they need plenty to drink. Most songbirds 'sip and tilt' to drink – scooping water into their beaks, tilting their heads back and letting gravity do the work. Woodpigeons, though, can use their beaks like straws to suck up water. Woodpigeons enjoy water for bathing too – after summer rain they sometimes wash in fresh puddles and often stand with one wing raised whilst getting clean, like someone soaping up in the shower.

Woodpigeons also have another special skill. A stretchy pouch in the pigeon's throat is known as a crop. Both male and female pigeons can make 'milk' in their crops to feed their nestlings. Like the milk of mammals, it is rich in protein and full of goodness. As the young birds grow, the pigeons mix crop milk with grains and seed to make their toddlers pigeon-porridge.

GRAVEL AND GRIT

Woodpigeons eat grit to grind up food in their stomachs. Look out for Woodpigeons waddling by the roadside – they come off the fields to peck up the gravel.

BIRDS OF A FEATHER FLOCK TOGETHER

Large flocks of Woodpigeons are a common sight in rural areas. They often travel in flocks to feed on growing corn in fields, and later on ripening grain. In winter, Woodpigeons roost communally, and large numbers descend on fields of winter crops to eat the young shoots. Two or three sentinels keep watch whilst the flock feeds. If you can see the bird's white wing-bars it means they've lifted their wings, ready to take off at any moment. The white bars give the flock a clear visual warning of potential danger. A Woodpigeon's white wing-bars also stand out as it flies – perhaps helping a group stay together, as each bird can follow the 'tail lights' of the birds in front.

You might see a male Woodpigeon trying to impress a female by standing up tall and sticking out his chest. The male also shows off to the female (and proclaims his territory) with a roller-coaster airshow known as a display flight. He soars upwards, snaps his wings together with a clap, then glides down with tail spread wide.

LISTEN TO THE DAWN CHORUS

In late April and May, in the still air before sunrise, a chorus of birdsong heralds the morning. A dawn chorus sit spot (see page 86) is a special way to start the day.

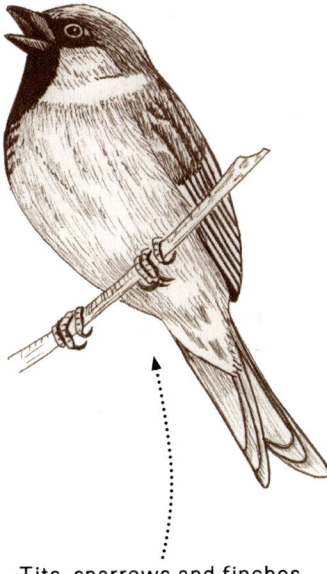

Birds with large eyes, such as Robins, Blackbirds and Song Thrushes, see well in the half-light, and often sing first. Listen for the early birds – the Blackbird with his rich, fluting song, and the Robin, with his cheerful, twiddly-scribbly warble. The Song Thrush repeats each phrase over and over in his loud clear voice.

As the sky lightens, more voices join the choir. The unhurried Woodpigeon joins in with his pleasant, comfortable coo.

Tits, sparrows and finches, whose smaller eyes don't see so well in the dim light, add their voices only once the sky is brighter. The two-note see-saw song of the Great Tit is easy to distinguish, as is the sound of the chirpy House Sparrow.

The first birds begin to sing about an hour before sunrise. You may be able to lie in bed and listen through open windows. If you live in the city, on a clear, still morning, get up early and head somewhere where there are trees and scrubby bushes – a back garden, square or park. Cemeteries are good listening spots because birds use the gravestones as songposts. Wear warm layers, take a flask and enjoy the choir.

COLLARED DOVE

Streptopelia decaocto

Black half-collar

Soft pale grey

In flight, the white tail tip is clear

Slim body.

Juveniles do not have a black collar.

VOICE

The Collared Dove sings from a high perch; a fencepost, rooftop or TV aerial. The song has three notes, with the emphasis on the middle note: *coo coooo coo*. Listen for the long reedy flight-call – *huuuww* – which often announces a Collared Dove's arrival as it comes in to land.

SIZE

31–33cm

WHERE

Farms, gardens, parks, towns.

NEST

Twig-platform near trunk of tree (often an evergreen).

EATS

Seeds of crops or grasses, berries.

MAY

82

COLLARED DOVE – THE OLD NAMES

From the Collared Dove's black half-collar: **Collared Dove**

Die Fernsehtaube – the way the Collared Dove calls from aerials
inspired the modern German name, television dove

RECORD BREAKING

The Collared Dove is one of our commonest British garden birds, so it's incredible that before the 1950s not a single Collared Dove lived wild in Britain. Originally from Asia, Collared Doves flew west and spread across Europe, expanding their range faster than any other species recorded in the last hundred years. In 1955, a pair arrived on the coast of Norfolk and nested in a garden. Within the next 20 years, Collared Doves could be found in any garden in Britain.

LOVEY DOVEY

Collared Doves are often seen in pairs. Watch them sitting side by side on rooftops, preening each other with affectionate bill-kisses and tender head-nuzzling. Like all pigeons, Collared Doves mate for life, forming devoted pair-bonds.

In many ancient cultures, including Sumerian, Assyrian, Greek and Roman, doves were associated with goddesses of love, Inanna, Ishtar, Aphrodite and Venus. No wonder 'billing and cooing' has come to mean behaving like a couple of lovebirds.

Watch for the male Collared Dove's courtship display: with flapping wings the bird soars upwards, then glides downhill (sometimes in a spiral), often coming in to land with a final flourish – a triumphant purring call.

FERAL PIGEON

Columba livia

Feral Pigeons can have many different colours and patterns, including pale grey, dark grey, brown and pure white – each individual is unique.

VOICE

A Feral Pigeon's song is a continuous bubbling *crroooo*, a familiar sound in town centres. When one pigeon returns to the nest after finding food, it is welcomed by the other pigeons' crooning greeting.

SIZE

31–34cm

WHERE

Towns, cities.

NEST

On ledge.

EATS

Buds, seeds, cereals, leftovers.

FERAL PIGEON – THE OLD NAMES

From Latin *pipio*, to chirp or peep: **Pigeon**

Common Pigeon/Street Pigeon/City Dove

WILD ANCESTRY

In ancient times, domestic pigeons were bred for food from the wild Rock Dove. The ancestors of today's Feral Pigeons were escapees from medieval dovecotes, returned to the wild.

See Feral Pigeons feeding on scraps and crumbs anywhere people eat lunch outside. Large populations live in city squares and town centres; look for them at railway and bus stations, in gutters and on rafters and ledges and roofs of buildings. Being so used to humans, they're often tame and happy to be fed by hand – just as 'The Bird Woman' feeds the birds on the steps of St Paul's Cathedral in *Mary Poppins*.

COMMUNITY GATHERING

Feral Pigeons like the protection of a roof overhead – an echo of the rocky overhangs that shelter wild Rock Doves. They also like to be part of a community – often dozens of birds share the same nesting site.

Look up, and you might spot Feral Pigeons nesting under bridges or along rafters and ledges of buildings. Their flimsy nests often include drinking straws and bits of wire – the urban equivalent of twigs and sticks.

Look for them in the morning, when they congregate on south-east facing roofs to bathe in the first rays of the sun.

BOBBING AND BOWING

The constant supply of food and warmth in cities means Feral Pigeons breed all year round. Watch the male birds strutting around the females, bowing and puffing out their chests to show off the metallic lustre of their neck feathers and fanning their tails behind them like royal trains.

Pigeons usually have two white eggs, hence the saying 'a pigeon's pair' – a family of two children. Pigeon parents take turns to sit on their eggs to keep them warm, the male by day and the female by night.

PIGEON POST

With a strong homing instinct, an extraordinary memory and an ability to sense the Earth's magnetic field, pigeons can travel across continents to return home to a specific street. For thousands of years, humans have used pigeons' skills to communicate over long distances – important messages were written on tiny rolls of parchment or paper and tied to the bird's leg. Ancient Egyptians used pigeon post to announce the coronation of Ramses III, and Romans sent news of chariot race winners by pigeon. In both world wars, pigeons carried messages for the British army, saving many lives. Royal Air Force aeroplanes often carried a pigeon on board, to fetch help in case of an emergency. In 1942, a plane was shot down and crashed into the freezing waters of the North Sea. The plane's pigeon flew over 120 miles, arriving in England the next morning oil-stained and exhausted. The rescue mission launched saved the lives of the whole crew, and she became the first creature to receive the Dicken Medal – the animal equivalent of the Victoria Cross. In the Second World War, 32 pigeons were honoured with the Dicken Medal.

THE WHITE DOVE

The words pigeon and dove are different words for the same avian family, Columbidae. Feral Pigeons come in many colours, including pure white, with pure white pigeons being known as 'white doves'. A pigeon's ability to carry a message informs the Christian symbolism of the white dove. In Catholic gospel the bird is a symbol of the Holy Spirit. Saints were often depicted with a white dove whispering into their ear – a messenger of divine inspiration. In the story of Noah's Ark, Noah knew the flood had receded when the dove returned with an olive branch – a sign of God's forgiveness. Even today, the white dove is known as a bird of peace.

In Scotland and northern England, a white dove settling on a house or coming to the window was seen as a sign of peace to a parting soul. In Donegal, it was said that the soul of a child, or a person without sin, takes the form of a white dove.

Even in Celtic times, doves were seen as birds of peace, love and harmony, and were sometimes offered at shrines or sacred springs. At a sanctuary in Gaul, carved images of doves were offered. At healing springs in Trier, in Germany, there were depictions of children holding offerings of doves. Perhaps these birds of peace were given in thanks for health because peace of mind is so related to health of body.

SIT SPOT

A simple and profound practice for connecting with birdlife, and all life.

- You can enjoy a 'sit spot' anywhere in nature. Find a comfortable place to sit, somewhere you can easily get to, such as your back garden or a local park.
- Move slowly and quietly to your spot, to avoid disturbing the birds and to quiet your own mind.
- Sit.
- Bring your awareness to your body. Allow your muscles to relax. Allow your breathing to slow.
- Take a few minutes to focus on each of your five senses.
- You could start with your sense of touch. Notice how your body feels upon the ground, or against the tree. Feel the air on your skin – is it warm, cool, dry, damp?
- Listen. What can you hear? Birds? Wind? Traffic? What's the quietest sound you can hear? How many sounds can you hear at the same time?
- Allow your gaze to soften, and include your peripheral vision.
- Focus on your senses of taste and smell. Is the air sweet? How does the earth smell?
- Tune in to whatever is unfolding around you. Sit for 5 to 20 minutes.
- Revisit the same sit spot, daily if you can. Notice what changes with the seasons, the weather and the time of day.

'LOVEY-DOVEY'
A FABLE FROM THE PANCHATANTRA

A story that celebrates the dove family's pair-bonding and community spirit

Ringdove and her mate were a devoted pair. They sat together side by side, cuddling and cooing. In the spring, they worked together to build their nest. In the summer, they took turns to warm their eggs and feed their chicks. In autumn, they joined the flock in the farmer's field, feasting on golden grain. And on winter evenings, they settled down to sleep alongside the flock, in the branches of the old oak tree, in the little patch of trees at the edge of the farmer's field.

But there was one person who wasn't so happy with the way the doves worked together. The farmer. And one day, the farmer set a trap. He hung a net between the branches of the trees. He took the string of the net in his hand. He hid himself, and he waited. And when the doves gathered together to feed on the winter kale, plodding and pecking and sticking close together, the farmer let go.

Down fell the net. The doves struggled and strained, but it was no good. They were trapped. The farmer rubbed his hands and licked his lips. 'Pigeon pie tonight!' Amidst all the flapping and flailing, Mrs Dove knew just what to do. 'Keep cool, dooo, keep cool,' she called. 'We feed together. We roost together. We can escape – together.'

So, together, they flapped their wings as one. Up, up, up into the air they rose, and there was nothing the farmer could do but clench his fists and curse. 'Pesky pigeons!'

Higher and higher they rose, flying as one, away to a far-off clearing. They were free from the farmer. But they were still stuck in the net.

Perched high on the lightning-twisted tip of a lonely oak was a crow. *Kraar! Kraar! Kraar!* A solitary crow, watching the flock of doves. He'd seen the way they worked together, and he was curious. Most of the time, he was alone. He liked his own company. 'I can help you,' said the crow. 'I know a little mouse who lives in the hedgerow. He may be small but his teeth are strong.'

Squeak-squeak! The mouse nibbled and gnawed the net with his sharp little teeth. The ropes of the net frayed and loosened, stretched and snapped ... and with a flutter of feathers and whistling wings, a rush of air, the doves flew free.

The lonely crow perched on his post and considered. He heard the *chack ... chack* of his cousins, the Jackdaws, calling to one another. He watched the Rooks patterning the evening sky, as they flew to roost. And he took wing, to join them.

The summer solstice is around 21st June – the time for midsummer magic. Nature is at its most luscious; the sun is warm, the air is soft. Wild grasses grow tall and tickly, and silky-skirted poppies dance in the fields. In the hedgerows, creamy clouds of elderflowers and delicate wild roses delight our senses. Summer evenings are scented with honeysuckle.

The summer skies are graced with welcome visitors – Swallows, House Martins and Swifts have flown thousands of miles to make their nests alongside our own homes. These visitors (known as 'migrants') arrived in spring and by the winter months will have flown to warmer lands – their fleeting presence makes their time with us all the more precious.

It is also a time for baby birds. By June, most garden birds are busy with growing families. Nests are loud with cheeping chicks and young birds take their first daring flights out into the world. Keep your eyes, and your ears, open for bird babies.

JUNE

Baby Birds

THE EGG

Female birds find the extra calcium needed to form an eggshell by eating tiny snail shells, or pieces of broken eggshell, tidied out of the nest by other new parents. Remember, if considering 'pest control' in our own gardens, that snails are vital to eggs.

You may find a broken eggshell, which a parent bird has removed from the nest. The size of the egg can give you a clue to the species and so can the time of year (since some birds only nest at certain times of year). The colour is a good clue too:

- **A blue or greenish egg**: Birds that nest in bushes or trees (like Blackbirds and Carrion Crows) generally have eggs in greenish (or blueish) shades – colours that are well camouflaged amidst greenery and dapples of skylight. Blue-green eggshells also act as sunscreen to protect the growing chicks inside from UV light.
- **A pale coloured egg**: Birds that nest in holes (like Great Tits and Tawny Owls) generally have white or pale blue eggs – the light colour makes them stand out in the dim light, so parent birds can see where they are. In nest-holes, eggs are already well hidden – they don't need to be camouflaged.
- **A brown or speckled egg**: Birds that nest on the earth, or out in the open (like Skylarks and Pheasants), generally lay brown or speckled eggs that blend in well with the colours of the ground. There is a theory, too, that darker eggshells keep growing chicks warmer, because dark colours absorb more sunlight than pale colours.

Female birds often lay at dawn, leaving time later in the day to feed. Most species lay several eggs, laying one egg a day – a bird's egg can weigh up to one-fifth of its own body weight, so a female couldn't carry a whole clutch of eggs in her belly at once or she'd be too heavy to fly.

Inside the eggs, the chicks can hear their parents' calls, and cheep back to them, and to each other. Sitting on the eggs to keep them warm (incubation) usually begins not after the first egg is laid, but once the whole clutch is complete, so that the eggs all hatch at roughly the same time.

IN THE NEST: NESTLINGS

Just as looking after human babies involves a lot of nappy-changing, bird babies also need help to keep clean. Chick poo comes in neat, jelly-like packages (known as 'faecal sacs') – the bird equivalent of a biodegradable nappy bag. When a parent brings food, it sometimes lingers on the edge of the nest, waiting for the chick to go 'bottom up', before taking the white blob away in its bill to drop it away from the nest. Just as toddlers learn to use a toilet, Swallow chicks, once they're five or six days old, learn to go over the nest-edge.

One clear sign that a bird is a youngster is a bright 'gape'. The inside and edges of many young birds' beaks are brightly coloured – Robin nestlings, for example, have a yolk-yellow gape. When a parent bird arrives at the nest the chicks reach up for food with loud calls and open beaks. The bright gape is a visual shout that says, 'Feed me! Me! Me! Me!' It also gives parents a clear target for aiming their food-drop. Newly hatched chicks will gape at anything (even predators!) but soon learn to lie low and keep quiet when they hear an alarm call. From dawn to dusk, parent birds search for food for their nestlings. The sound of loud insistent piping can help you spot a nest – watch from a distance (so you don't disturb the family) as parents fly back and forth with food.

OUT OF THE NEST: FLEDGLINGS

The comings and goings of busy parents and the chicks' noisy calls make the nest easier to spot, not just for people but also for predators. So parents encourage chicks to leave the nest as soon as they can, sometimes even before their feathers are fully formed – before they've mastered the art of flying.

Chicks sometimes need a bit of encouragement to leave the nest. Sometimes a parent offers food from outside the nest to lure the chicks to take off. Sometimes a mother will even practise tough love and push a chick out. The baby opens its wings and flutters its feathers to save itself from falling, and finds that it's flying. Most broods leave the nest (fledge) at dawn, hidden in the half-light. After a chick leaves the nest, it's known as a fledgling. Once fledged, most birds do not return to the nest.

Most small songbirds still depend on their parents for food for a few weeks after leaving the nest. In June and July, thousands of fledglings are being taught how to fly and how to find food. This offers a special chance to spot young birds, as fledglings come out into the open asking their parents for food. Chicks signal their hunger with loud persistent squeaking and by opening their mouths wide, showing their still-bright gape. All parent songbirds instinctively respond to the sight of *any* chick's open gape by fetching food. House Sparrows, Robins and Wrens, Blue Tits and Great Tits, Blackbirds and Starlings have all been recorded feeding chicks of a different species. A bereaved pair of Robins was observed feeding baby Song Thrushes, for example, and a male Robin brought up a brood of orphaned Blackbirds. Most fledglings retain a hint of a gape. Fledglings often look rather fuzzy round the edges too, having used their energy to grow, rather than perfect their feathers. Look out for fledglings with fluffy feathers, stubby tails and clumsy flight. Like toddlers learning to walk, fledglings learning to fly sometimes bump into things and have unexpected crash landings.

Fledglings often wander to the edge of their parents' territory, spread out slightly from their siblings – they're within hearing of the family, but there's less risk of a predator making an easy meal of the whole brood. Usually, a family keeps broadly together, with both parents feeding their babies – watching their parents find food helps the youngsters learn. Some fledglings (such as Song Thrushes, Blackbirds, House Sparrows and Robins) are at first looked after by both parents, and then left in the care of their father, whilst the mother tends to their next clutch of eggs.

If You Find a Fledgling

Leave a stranded fledgling be (and take your pet cat indoors). A parent is usually close by, waiting for you to leave, so they can feed the chick, or lead it to safety. Do not try to return a young bird to the nest – the disturbance could make parents abandon the nest. Only move a fledgling if it's in danger, such as on a busy road, and only move it to within hearing distance of where you found it.

JUVENILES

By the end of July, the garden is aflutter with newly independent birds, now known as juveniles. Once their wing and tail feathers are fully grown, juveniles look like adults in size and shape, though many young birds keep their duller colours, which camouflage them from predators. You can also tell a juvenile by its behaviour – if you see one bird opening its beak and fluttering its wings at another, it's probably a fledgling asking a parent for food.

In August, juveniles disperse, leaving their parents' territory to explore the local area. Some birds, such as Starlings, gather in flocks when they leave their parents, and roam the area together, looking out for future homes and nest-sites. Sometimes young birds of different species flock together – look out for mixed flocks of young tits roving the hedgerows.

Spotting Fledglings

Blackbird and Robin fledglings are both bold, and so easier to spot.

Baby Blackbird

Baby Blackbirds usually leave the nest before they can fly, so look out for them hopping around trees and bushes. Listen for begging calls from April to August. You may see a baby Blackbird following a parent across the lawn, squeaking for food, and hopping up close whenever a worm is found.

You can tell a young male even after it's grown black feathers, because, in his first winter, his beak is still dark, not yellow. Even the year after fledging, some of his wing feathers are still a rich brown colour.

Baby Robin

Look out for baby Robins near the shelter of garden shrubs, rather than in the middle of a lawn. You might see them fluttering their wings at the edge of a border whilst a parent forages in the under-growth.

Russet brown (similar to its mother, though adult females are more chocolate brown)

Downy feathers can make a young Blackbird look bigger than its parents

Breast is more speckled than adult female's

BABY BLACKBIRD

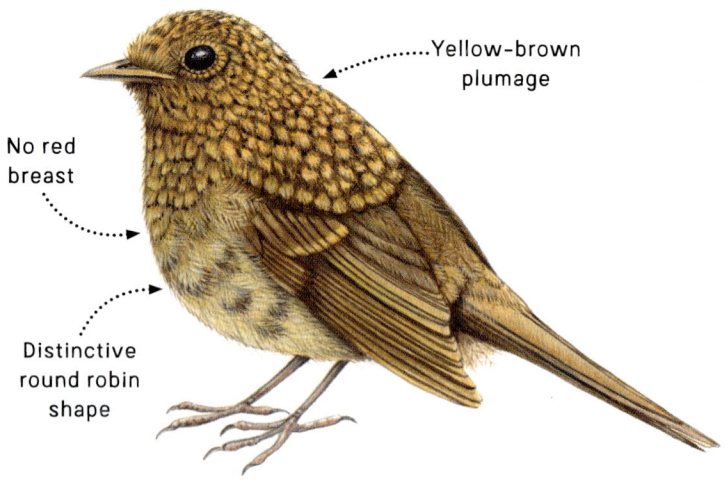

Yellow-brown plumage

No red breast

Distinctive round robin shape

BABY ROBIN

SWALLOW

Hirundo rustica

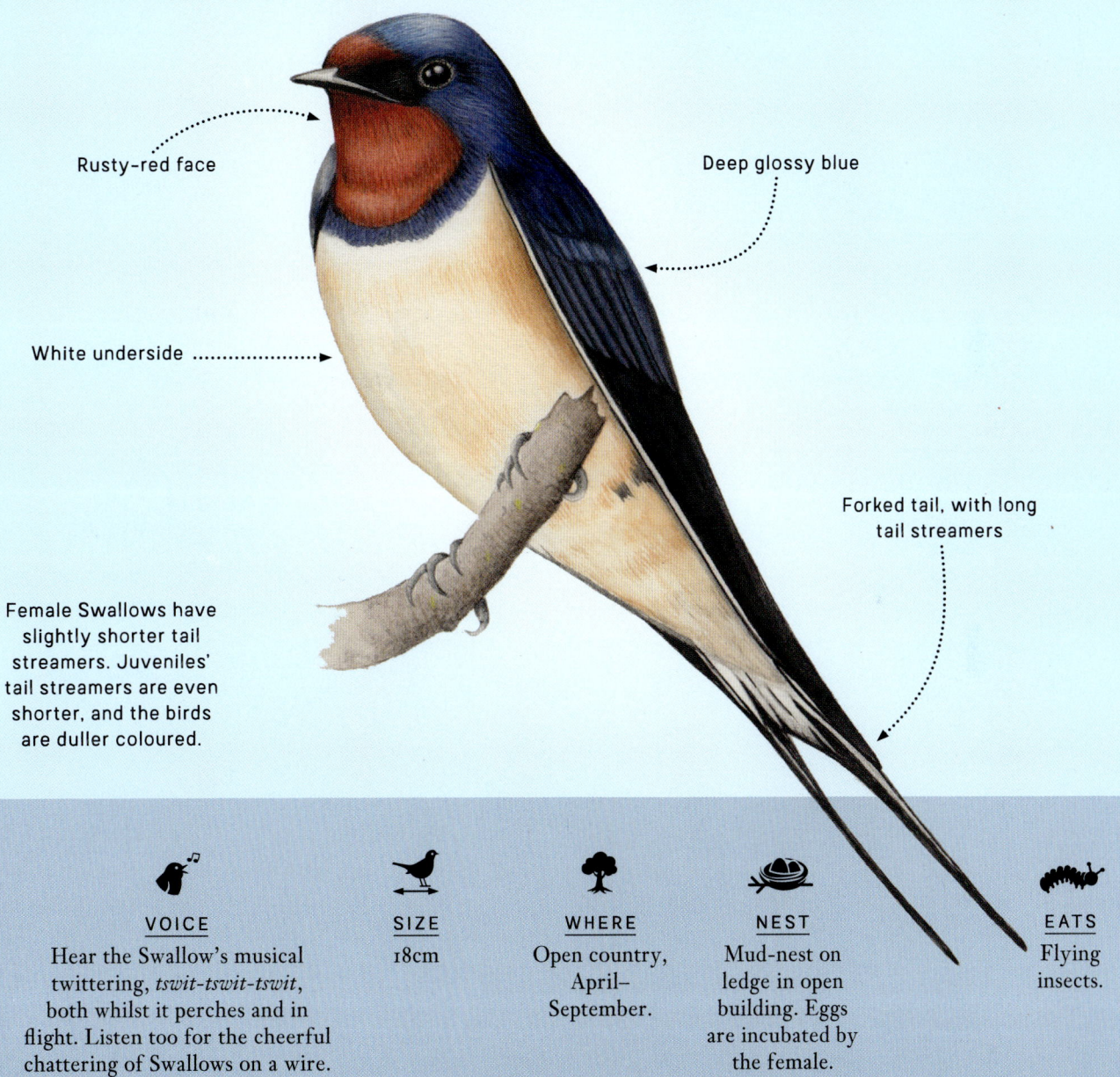

Rusty-red face

Deep glossy blue

White underside

Forked tail, with long tail streamers

Female Swallows have slightly shorter tail streamers. Juveniles' tail streamers are even shorter, and the birds are duller coloured.

VOICE
Hear the Swallow's musical twittering, *tswit-tswit-tswit*, both whilst it perches and in flight. Listen too for the cheerful chattering of Swallows on a wire.

SIZE
18cm

WHERE
Open country, April–September.

NEST
Mud-nest on ledge in open building. Eggs are incubated by the female.

EATS
Flying insects.

SWALLOW OF SUMMER, BRINGER OF SPRING

Fly away, fly away, over the sea,
Sun-loving swallow, for summer is done.
Come again, come again, come back to me,
Bringing the Summer and bringing the Sun.

'THE SWALLOW', CHRISTINA ROSSETTI

Many of our feathered friends, though born in Britain, spend most of the year in warmer lands further south. The Swallow that skims the summer skies has flown thousands of miles to reach northern shores, to find a place to nest and bring up its young.

The Swallow is one of the first of our summer visitors to arrive, from early April onwards. The birds reach our shores in waves – the first arrivals are males returning home, usually older, more experienced males, who head straight to established breeding sites (often those used for generations) to rest and feed for a few days, before the females arrive to join them.

A 2,500-year-old Greek vase shows a man and two boys looking up at the sky. The figures have words coming out of their mouths. The first boy calls, 'Look, there's a swallow.' The man replies, 'By Herakles, so there is!' The second boy raises his arms in greeting, crying, 'There she goes. Spring has come!'

Because Swallows seem to arrive towing the warm weather behind them, they are revered far and wide as bringers of spring, associated with hope and renewal, an essential part of the summer sky. As Ted Hughes says:

When the swallow snips the string
that holds the world in
And the ring-dove claps and nearly
loops-the-loop
You just can't count everything that
follows in a tumble
Like a whole circus tumbling through a hoop

FROM 'APRIL BIRTHDAY'

The Romans believed Swallows flew to us from paradise, bringing with them warmth to the Earth. The sight of the first Swallow was celebrated in many lands. The ancient Greeks held a festival to honour its arrival on the island of Rhodes, a tradition still practised until the beginning of the twentieth century. Bands of children paraded the street carrying a pole topped with a wooden Swallow, and welcomed the bird with a song:

The swallow comes! She comes, she brings
Glad days and hours upon her wings.

FROM THE ENGLISH VERSION
OF THE FOLK SONG, TRANSLATED BY
COUNTESS MARTINENGO CESARESCO

In Albania too, children sang a song to welcome the Swallow:

O swallow, little swallow,
Welcome as you now appear!
The snow has gone from off our mountains,
Winter's ended, spring is here.
Build your nest beneath my window,
Wake me with your morning song.
And we'll give you and your nestlings
Food for all the summer long.

In Russia too, songs were composed to celebrate the Swallow's return after the long dark winter, and in Westphalia, farmers and their families would wait at the gates to welcome the birds, throwing open the barn doors so they could enter. In German villages, town guards announced the arrival of Swallows with trumpet fanfares. In Brittany, Swallows brought good fortune to homes that harboured them, and children called up to the birds, 'come and build your nest in my … window!'

In Cornwall, it was the custom to jump in the air at the sight of the first Swallow. And in County Clare, in Ireland, when you see the first Swallow, you can make a wish.

The saying 'one swallow does not make a summer' comes from Aesop's fable *The Spend-thrift and the Swallow*. Though our minds know it's true, still, our hearts lift at the sight of the first Swallow of spring.

THE MYSTERY OF MIGRATION

Whilst in Britain, Swallows nest and bring up their babies. By late August, they will be leaving us again, flying away with their young. Where Swallows went was for many decades a great mystery. Writers from Aristotle to the eighteenth-century naturalist Gilbert White suggested that Swallows hibernated in the bottom of ponds, in the mud.

The idea probably arose from seeing Swallows sink down to roost in reed beds.

We now know that British Swallows spend their winter in sub-Saharan Africa, where the skies dance with bugs, and they follow in the wake of herds of elephants and buffalo, who throw up clouds of insects in their footsteps – a reality no less wondrous than Aristotle's idea.

Migrating Swallows can travel 200 miles a day, journeying over 5,000 miles to reach southern Africa. They cross the English Channel, follow the coast of France and fly over Spain, stopping from time to time to rest amidst the orange and lemon trees. Over the ocean, and the Rock of Gibraltar, they go, and rest again in the date palms on the shores of Morocco, before flying over the snow-capped peaks of North Africa. Their journey often crosses the Sahara (a waterless desert 930 miles wide) in one go. They fly over the tropical forests of central Africa, and reach southern Africa some time in December.

At the end of the African summer, they make the journey again, in reverse – a round trip of 10,000 miles! Their journey to our shores, though, takes only four or five weeks, with the urge to nest and breed speeding the birds on. After their epic journey, they see again the white cliffs and green fields of England, and return to the very same landscape of village ponds, thatched cottages and beamed barns where they were born.

But *how* do they find their way on such an epic journey? Much is still a mystery.

A legend from Estonia tells that the Milky Way is the bridal veil of the daughter of Ukko, God of the Sky, and that it is she that directs migrating birds. A goddess with a scarf of stars, guiding the birds! The vivid image rings with truth as well as beauty – many small birds do fly by night, using the stars to navigate. Young birds are not born with a map of the night sky in their mind's eye – they learn the patterns and movements of the stars.

Yet Swallows migrate by day, so they can feed on insects on the way. It is thought that using the position of the sun to find south (for example, keeping the setting sun on their right – in the west) helps migrating birds travel south.

Flying by day also means they can use landmarks to navigate, such as rivers, forests and coastlines. Some Swallows, for example, follow the west coast of Africa to find their way around the Sahara. It is thought that Swallows also detect wind currents, air pressure, infrasound and polarised light. Like many birds, Swallows are also sensitive to the Earth's magnetic field, having a kind of 'internal compass' to help them navigate (especially useful when the sun is hidden by clouds). Imagine being so sensitive to the invisible workings of the world!

SWALLOW TAIL

The swallows flew in the curves of an eight
Above the river-gleam
In the wet June's last beam:
Like little crossbows animate

FROM 'OVERLOOKING THE RIVER STOUR',
THOMAS HARDY

The Swallow's long tail streamers give the bird lift and, when spread, help the bird slow down and change direction with agile ease. They are masters of manoeuvring, able to 'brake' and 'turn tail' with fluid speed. In experiments, scientists attached long tail streamers to House Martins and found it improved their ability to zip and zigzag through a maze.

The long, graceful tail streamers of the male Swallow also help them to attract a mate. The male sings vigorously, and glides slowly overhead, fluttering around the female and fanning out his tail.

Traditional tales offer many explanations for how the Swallow gained such a distinctive tail. In Belgian lore, the Swallow tried to bring humans fire from heaven, carrying the flame on his tail. The fire burnt away the bird's central feathers (and the flame flew back to heaven).

In Siberia, the Buryat peoples told that the Swallow stole fire from heaven, and Tengri, the sky god, fired an arrow at the bird, hitting its tail and removing the central feathers.

A French tale tells that on Noah's Ark, the Swallow tricked the snake into eating mosquitos, instead of the flesh of his human friends. The angry snake tried to catch the bird, but only managed to snap up a mouthful of tail feathers, as Swallow flew free.

SWALLOW'S NEST

Swallows once nested in caves. But at least since Egyptian times, they've nested in human buildings, especially those near open farmland. Traditional dairy farms are perfect Swallow habitats. Farm animals leave dung, which attracts insects. Cows' hooves churn up mud, needed for nest building, and cows don't crop the grass too short, so insects flourish in the pastures. Chicken feathers, floating on the breeze, make a snug nest lining.

Old outhouses offer Swallows perfect nesting sites. The roof gives protection from the elements and predators, and old beams provide a stable base for a nest – a mud-nest sticks well to rough wood.

To create a nest, both parent birds gather mud from the edges of ponds or puddles, or even from rainwater gathered in a cow's hoof-print. A German proverb celebrates this teamwork: 'When both swallows fetch mud, the nest is soon made.' The birds collect the wet mud in the morning, so it has time to dry throughout the day. They make their nest on top of a beam or ledge, building from the bottom up by placing balls of mud on top of each other. Each mud pellet is pressed into place using beak and tongue. The bird vibrates its head to

squish down the mud, squashing out airspaces that could cause cracks when it dries. The mud is mixed with grass, straw or horsehair, and formed into an open cup shape, about the size of a grapefruit. The cup is lined with dry grass, then warm feathers. The female gradually removes this feather bed as her chicks grow, to stop her babies getting too hot. Look beneath the beams of old farm buildings and you may find the feather blankets from a baby Swallow's bed.

Making a new nest takes more than a thousand mud-collecting trips. So, Swallows often reuse their old nest, plastering cracks and strengthening crumbling edges with fresh mud, and adding a new feather lining. A nest can last for 10–15 years, and be reused every year.

SWALLOWS ON A WIRE

... And gathering swallows twitter in the skies.
CLOSING LINE OF 'TO AUTUMN', JOHN KEATS

In the golden late summer, before their long journey south, Swallows and House Martins flock together. See them perched in a line on high wires, strung like notes upon a stave. They preen their feathers and twitter and chatter, lift off, fly around and land again – they seem restless, excited even. Their first journeys south are short stopovers, as they rest in reed beds en route, to fuel up on September swarms of midges and roost in the safety of reeds and willows. Roosting in water plants protects the birds from ground predators – it's impossible to sneak up and strike in silence in such splashy surroundings.

In autumn, when insect food is scarce elsewhere, reed beds can shelter thousands of birds. Visit a reed bed at dusk and you may be lucky enough to see Swallows sweeping down to roost.

In this way, Swallows move south step by step, until they come to the English Channel in September or October (or perhaps, for a few stragglers, even in November) and leave our shores, until next year.

THE SWALLOW – A GIFT

The swallow and the swift
Are God Almighty's gifts.
HEREFORDSHIRE RHYME

Having swallows as next-door neighbours is good luck – the idea spans the globe, just like the Swallow. It was known by the Romans and in ancient China, as well as in England and Ireland. Chinese buildings even had special ledges added, to encourage the birds to nest. As the Croatian proverb says, 'Swallows turn a house into a home.'

Destroying a Swallow's nest is, of course, bad luck. In France, it was thought that if a Swallow's nest was removed from a stable, one of the animals within would fall lame. In both English and Irish folklore, harming a Swallow's nest spoils the cow's milk. English folklore warns that such destruction can also stop hens laying and bring rain down on crops for a full month. Country folk the world over knew how important it was to treat the Swallow with care.

FEATHERS OF RUSSET RED

Whilst older tales link the Swallow's red feathers with fire, Christian legends tell that the Swallow's feathers turned red when the bird helped Christ on the cross. Russian folklore says the Swallow tried to remove the nails. In Tuscany it was said that the Swallow removed a thorn from the crown. In Portugal, they said Swallows wiped the blood from Christ's wounds. A related belief, from Brittany, was that Swallows always arrived in time for Good Friday.

SWALLOW WEATHER LORE

In England, France and Ireland, Swallows flying high was understood to be a sign of good weather. Peoples of the past had more direct contact with nature than modern city-dwellers – their observations, which crystalised into gems of folklore, often hold truth. In warm, settled conditions, the insects Swallows eat are carried upwards on rising thermals. When the air is humid, and rain is more likely, insects fly low, and Swallows follow. Hence:

When Swallows fly low,
A storm will blow.
ITALIAN SAYING

Swallows skimming o'er the plain
Are sure messengers of rain.
FRENCH SAYING

'WHY SWALLOW MAKES A MUD-NEST' A MAYAN FOLKTALE

A story that celebrates the Swallow's mud-nest

Long, long ago, when the Earth was young and the birds were new, each little bird was learning how to make a nest.

In the sunny lands of the south, Flycatcher was making a nest safe in a hole, in a tree.

Oriole was weaving a basket – it hung like a hammock beneath a branch, swinging gently in the wind.

But Mr and Mrs Swallow were not making a nest at all. They were enjoying the sunny skies. The days were long and light. The sky was dancing with tiny, winged things – a feast of flying insects.

They loved to swoop and swerve, dart and curve, sweeping low over land and dipping down over water. They twittered to one another and flitted to and fro, skimming the skies.

When autumn came to the south lands, the sun-loving swallows began to think of journeying north. Because, when summer is ending on the south side of the Earth, on the north side of the Earth, summer is just beginning.

Every year, when summer ended in the south, the swallows flew north. They stayed in the lands of the north all through the summer, and when autumn came around, off they flew again, south to where summer was just beginning.

This year, the swallows were ready to start a family. They were hoping that, once they arrived in the north lands, they could make a nest and lay some eggs.

'We'd better learn how to make a nest, before we go,' said Mrs Swallow.

So off they flew to see Flycatcher. 'Please, dear Flycatcher, will you teach us how to make a nest?'

But Flycatcher had already made her nest, and it was full of hungry babies, all piping to be fed – she was much too busy to show the Swallows what to do.

So the Swallows flew to Oriole. 'Please, dear Oriole, will you teach us how to make a nest?'

But Oriole too had made her nest already, and it was full of hungry chicks – she was busy with her babies, and she shook her head.

Each and every bird the Swallows asked shook their head – they were all busy feeding their own families.

'Oh me, oh my,' twittered Mrs Swallow. 'We've left it too late. We'll never learn to make a nest, and I'll have nowhere to lay my eggs.' And she burst into tears.

Mrs Swallow perched at the edge of the water-hole, and she cried and cried. But between her sobs, she heard another sound. A high, thin cry. 'Help! Help! Please, somebody, help!'

The voice was coming from deep down inside the waterhole. Mrs Swallow peered over the edge. There was a little wasp, clinging onto a leaf. The leaf was twisting and turning in the deep water. 'Help!' cried the wasp, 'I fell in the well! I can't swim!'

'Hold on!' called Mrs Swallow, and down she swept.

And back she came with the dripping leaf in her beak, and the little wasp clinging on tight. 'Oh, thank you,' gasped the wasp. 'You saved my life!'

'You're welcome,' said Mrs Swallow, 'I'm happy to help.' She gave a great sniff.

'Then why are you crying?' said the wasp. And Mrs Swallow told her the whole story.

'Well, that's no problem,' said the wasp. 'I can show you how to make a nest.'

The little creature was a mud-wasp – she made her nest from mud. She showed the swallow just what to do. First the wasp found a protected place. Then she scooped up some good sticky mud. She plastered the mud into place and pressed it into shape, humming as she worked. Soon enough, she'd made a nest of good strong earth.

'Why, thank you, thank you!' twittered Mrs Swallow. 'That's just what I need to know.'

So, off the swallows flew, on their long journey, over lush trees and dry sands, high mountains and deep seas. Every day they flew hundreds of miles, until at last they arrived in the lands of the north, and saw familiar fields and farms beneath them.

They flew to a farm, a friendly place they knew, with clucking hens and quacking ducks and a big old barn. Home.

Straight away, they began to make their nest. They found a protected place inside the barn, up on a beam. They swooped down and scooped up some good sticky mud, from the edge of the pond. They plastered the mud into place. They made a neat mud cup. They swept the skies, catching fluffy feathers that floated on the breeze – the feathers Speckled Hen didn't need anymore – and they lined their nest, soft and warm.

Mrs Swallow laid four fine eggs. Soon enough, the big old barn was filled with the peeping and piping of babies. Little balls of fluffy feathers, black and white with wide mouths and yellow beaks.

And from that day to this day, Swallows always make their nests from mud, just as the little wasp taught them, long, long ago.

WELCOME THE SWALLOWS

Weave together inspiration from ancient customs and your own appreciation for the Swallow to create your own family tradition, to welcome the Swallows home.

You could make a Swallow on a stick, like children in Rhodes once did:

1 Cut out two pieces of cardboard, using the templates on page 190.
2 On each piece, where the wing joins the body, fold the card over. When the Swallow is finished, the wings will stick out at either side of the body (like the wings on a paper aeroplane).
3 Hold the two pieces of card together, so you can see the side of each piece of card that will form the outside of your bird. Colour each piece of card on its 'outside' face – look at a Swallow picture (page 93) to help you. Colour both sides of each wing.
4 Find a good stick.
5 Glue the two pieces of card together, with the end of the stick between them. Use a few staples if necessary.
6 When you move the stick, the Swallow flies!
7 You could use your Swallow stick as a wand, to make wishes and sprinkle blessings upon all the places Swallows might need or enjoy.

If you know where Swallows have nested before, that's a good place to watch for their return. You could hold a welcoming party after you spot the first Swallow, or choose a set date to celebrate the Swallow (in England, 15th April was once known as 'Swallow Day'). You could even write a song or a poem, and sing it to welcome the Swallows.

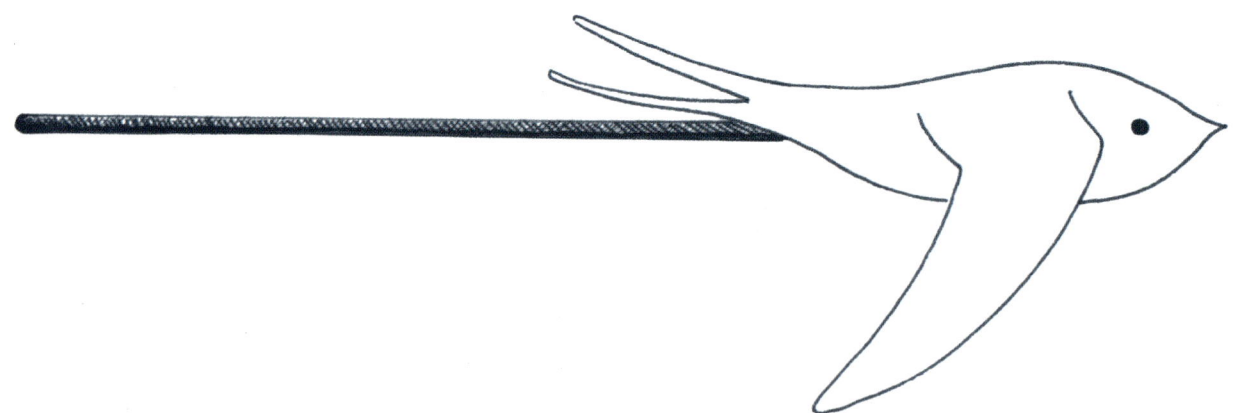

HOUSE MARTIN

Delichon urbicum

Glossy blue-black
back and cap

White patch
on rump seen
when flying

White leg
feathers

White underside

Juveniles have
duller colours and
shorter tails.

VOICE

Often many House Martin
voices are heard at once, twittering
and chirruping softly as they flutter
to and fro. A House Martin's
voice can sound quite buzzy, like
a scrambled telephone exchange.
Their call is a hard, dry *prrit*.

SIZE

12.5cm

WHERE

April–
September,
around houses
near farmland/
wetland.

NEST

Closed cup-nest,
under eaves.
Both birds
incubate the
eggs.

EATS

Flying insects.

HOUSE MARTIN – THE OLD NAMES

Named after St Martin – the oldest use of a Christian name for a bird

Martlet – diminutive of 'Martnet', short for *Martinet*, the French name for a Swift

Before the French name was used, the House Martin was known as a Swallow,
and distinguished by where it nested:
Eaves Swallow (Craven) • Window Swallow (Northumberland)

THE HOUSE MARTIN NEST – A MUD MASTERPIECE

Watch for House Martins arriving some time in the second half of April. For the first few days, they fly with Swallows, zooming over pools and river meadows for insects. Before humans built homes, House Martins nested on cliffs. Now, as their name suggests, they nest under the eaves of human houses.

The birds begin nesting around mid-May. House Martin nests rely on mud, so the birds often nest somewhere near a good mud source, such as a village pond. A pair collects beakfuls of mud from the edges of ponds, puddles or streams, and sometimes kneads together wormcasts found on the grass. This is one of the few times this aerial bird

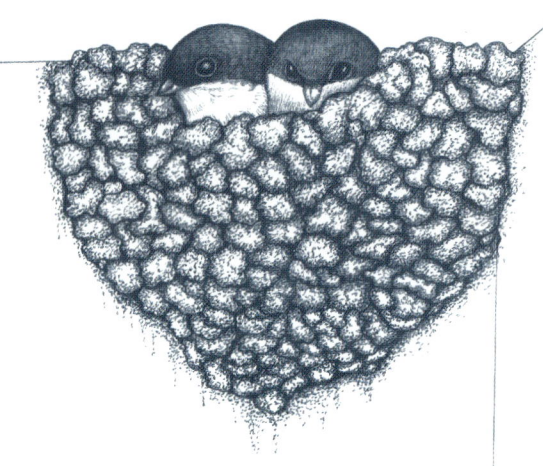

can be seen on the ground, waddling on its short, white-feathered legs. It takes around 2,500 beakfuls of mud, as well as feathers, grass and straw, caught in mid-air, to make the nest. The mud-nest is built against a wall, often under the eaves. The apex of a gable roof is a favourite nest-site, protected from the weather on three sides. The rounded nest has an entrance hole near the top – just below the roof of the house – so it's difficult for a predator to enter. If a nest is well sheltered from the weather, House Martins may reuse the same nest year after year (though not necessarily the same pair), repairing and restoring the old nest each year. Some ancient buildings, such as castles and churches, host colonies of House Martins, whose families have lived on the same stonework since the walls were first built.

House Martins are sociable birds, nesting together in a colony. The nests may be lined up along a wall, crowded together in a row, like a street of terraced houses.

Humans have been wise enough to realise that, like their cousins the Swallows, House Martins are blessed birds, and that to have them nesting in your eaves is good luck.

The martin and the swallow,
Are God Almighty's birds to hollow (hallow).
ENGLISH RHYME

YOUNG HOUSE MARTINS

Once chicks are hatched and ready to fly, House Martin parents have a special display to lure them from the nest, which they repeat over several days. One parent flies up to the nest, repeatedly hovering and calling in front of the entrance. If the young still sit tight, the female eventually lands at the nest entrance, calling loudly – a cheer of encouragement perhaps? Or perhaps a scolding telling-off!

Parents feed fledglings for several days, and the young birds return to the nest to roost, all cramming in together. Eleven House Martins were once counted sleeping together in a single nest – both parents and both their first and second brood of chicks. Fledglings may stay with the colony for a few weeks, and if their parents have a second brood, help to feed their younger brothers and sisters, before 'dispersing' to join flocks gathering together before migration.

BIRDS ON A WIRE

In autumn, watch for House Martins lining wires along with Swallows, as they gather together before flying to Africa. A wire is a perfect perch – giving a clear view all around. A mid-air perch also makes it easy to take off straight into fast flight.

House Martins generally migrate through France and Spain, though exactly where they go when they leave our shores is still unknown. Although over 290,000 birds have been ringed in Britain and Ireland, very few of these have been recovered – their winter ways are still a complete mystery. One theory is that the birds feed high above the uninhabited rainforests of Africa, where they are never seen by human eyes.

HOW TO HELP HOUSE MARTINS

Between 1970 and 2014, there was a 47 per cent decline in the number of House Martins in the United Kingdom, and in 2021, House Martins were moved to the 'red list' of Birds of Conservation Concern. Habitat loss has affected Martins and the insects they eat. Insect numbers are also affected by pesticides. Changing weather patterns can affect migration and lack of rain dries up the mud needed for nests.

Traditional styles of architecture, with open eaves, gave House Martins sheltered nest-sites. Later buildings had fewer suitable spaces for nesting Martins. Nests built on plastic surfaces, for example (as opposed to stone or wood), are much more likely to collapse.

There are many ways we can help:

- If your house doesn't have eaves, you could fix a weatherproof wooden board (at least 20cm deep) to create an overhang to offer shelter for House Martin nests.
- You could put up a pre-made terracotta House Martin nest, available from the RSPB.
- You could create a pond (page 75) in your garden, school grounds or local community green space, to attract insects for House Martins to eat and to provide mud for their nests.
- If the spring weather is dry, and a pond is not possible, even the youngest children will enjoy making a muddy puddle for House Martins.

SWIFT

Apus apus

Long pointed wings,
shaped like a bow

Short, forked
tail often seen
when flying

Pale chin

Dark sooty-brown
feathers – look black
when seen against
bright sky

Juveniles are blacker than
adults, with more white on
their throats and foreheads.

VOICE
The Swift's loud
piercing call, *screee*,
is unmistakable.

SIZE
16.5cm – shorter
body than the
Swallow, and
longer wings

WHERE
In the sky. April/
May–August.

NEST
In cavity. Both
birds build nest and
feed young.

EATS
Insects, spiders.

SWIFT – THE OLD NAMES

From the Swift's high speed: **Swift** • **Whip** (West Riding)

From its piercing call: **Screecher** (Hampshire) • **Screamer** (Sussex)
Squeaker/Squealer (Sussex) • **Screek** (Gloucestershire)

From its colour: **Black Swift** (Kirkcudbright) • **Black Screech** (Somerset)

The bird's wild cry and hurtling speed had negative associations for God-fearing Christians:
Devil Bird (Yorkshire) • **Swing Devil** (Northumberland) • **Devil's Screecher** (Devon)
Devil's Shrieker (Craven)

SWIFT AS A SWIFT

Fifteenth of May. Cherry blossom. The swifts
Materialize at the tip of a long scream
Of needle. 'Look! They're back! Look!'
And they're gone
On a steep
Controlled scream of skid …
… They've made it again,
Which means the globe's still working,
the Creation's
Still waking refreshed, our summer's
Still all to come

FROM 'SWIFTS', TED HUGHES

The Swift holds the record for fastest bird in level flight (almost 70mph, and 140mph when diving). The only predator fast enough to try catching a Swift is a Hobby, whose sweeping shape matches a Swift's aerodynamic curves.

Swifts cut quite a dash as they slice the sky. They're streamlined for speed, with long, narrow, swept-back wings, tiny tucked-in feet (like lifted aeroplane wheels) and a tapered, torpedo body.

The bird's neck is short, its head round, and its bill tiny. The spray of feathers in front of its eyes reduces glare, like the goggles worn by pilots in the 1930s and 1940s. Poet Edward Thomas described the form and flight of the Swift perfectly: 'as if the bow had flown off with the arrow.'

As well as flying fast, Swifts fly high. They can climb to an altitude of 3,000m, where the air is thin and cold, and oxygen is low, spiralling on stiff wings to catnap or to feed on insects carried high by currents of air.

THE SWIFT OF CEASELESS SUMMER

Swifts are birds of still, summer skies. Wind and clouds make it difficult for Swifts to hunt, as in long, cool, rainy periods, insects are scarce. A 'depression' causes the number of airborne insects to drop – a sign Swifts can read to sense weather coming from 300 miles away. Not only can Swifts sense rain coming, but they can also out-fly it. Birds that aren't busy breeding may travel over 100 miles to avoid heavy rain – their capacity for speed means that if weather is affecting insect numbers, Swifts in southern Britain just zip over the Channel to feed in France, while Swifts in west Scotland nip over the sea to Ireland.

Like Swallows and House Martins, Swifts undertake an epic journey of migration to Africa. The first to leave our shores are new fledglings, immature 'yearlings' and the year's non-breeders. Next go breeding males, and lastly breeding females.

LIFE ON THE WING

The Swift is a true bird of the sky. Swifts rest in the air, drink in the air and feed in the air.

Swifts eat over 500 kinds of airborne insects, as well as spiders that parachute through the sky on silken webs. Though their bills are tiny, their gapes are wide, and surrounded by bristles that funnel food into their mouths.

But they're not just catching whatever happens to be passing, like a trawler ship. A Swift will catch a stingless drone honeybee, but leave the stinging worker bees.

The birds drink on the wing, catching raindrops in mid-air, drinking from clouds, or swooping down to scoop up lake water. To shower, a Swift dives through a cloud, or vibrates its wings in a rainstorm.

Swifts even sleep on the wing. They are expert gliders, covering great distances with hardly a wing-beat. In the evening, they ride high on thermals to roost in the sky, then spend the night circling slowly down, dozing with one eye open whilst the other side of their brain is switched off. By dawn, they're back to the insect layer, ready for breakfast.

The Swift's scientific name, *Apus apus*, comes from the Greek for 'without feet', as the birds never perch on a wire or branch or on the ground. The hooked toes on their tiny feet all point forward – they cannot grip a horizontal perch. Still, they can cling to a rough vertical surface, like a wall, when landing at a nest-site, and shuffle about inside.

The only time Swifts touch down is to breed – eggs, unlike birds, need more than open sky to hold them. Swifts find a mate on the wing and gather all their nest material in the air, catching bits blowing on the breeze – a feather, perhaps, or the papery case of a leaf bud. These are glued together with their own saliva to make a nest, in a gap in a building, such as the crack in the stonework under a roof. Whilst their young need them, Swift parents swap their sky-beds for a more grounded bedroom, and sleep in the nest.

A SCREAMING PARTY

On fine summer evenings, watch for parties of Swifts swinging through the sky or chasing over rooftops with high wild cries. 'Screaming parties' even career round city streets.

Young swifts, not yet of breeding age, fly round the territory, learning the landmarks, and checking out which breeding holes are in use and which might be free. They sometimes fly right up to a nest-site, brake suddenly and knock the wall with a

wing (known as 'banging') – an escapade that brings the nesting bird inside screeching to the entrance, to confirm without doubt that the nest *is* in use.

SWIFT'S NEST

Though Swifts once nested in natural cavities, such as caves and tree hollows, they've lived with us in human homes since (at least) Sumerian times, 7,000 years ago. One of the earliest records of a specific colony is of Swifts living in the Western Wall in Jerusalem. Still today, Swifts whizz above the worshippers in the Holy City. In 2012, the mayor of Jerusalem held a ceremony to welcome the Swifts back to the Holy Land after their long 'pilgrimage'.

Today, Swifts breed almost exclusively in holes in human buildings – gaps and cracks in walls and windowsills, and spaces underneath the eaves – in homes, towers and churches. Swifts find it easier to nest in buildings made before the Second World War – the Modernist architecture that followed features sharper lines and offers fewer of the hollows they need. Town planners, developers and architects can all help Swifts by designing buildings that suit bird families, as well as human families. Including 'swift bricks', for example, is a cheap and simple way to offer the birds a home in new-build houses.

When Swifts arrive in Britain, they head straight to their nest-sites, usually nesting in a colony of several pairs. Pairs stay together for life and often reuse the same site. A Swift will greet its mate by showing its white throat – a display that shows its peaceful intentions. A pair preens each other tenderly, nibbling their partner's face and neck feathers.

Both parent birds build the nest and take turns to warm their eggs and feed their chicks – each flying up to 500 miles a day to search the skies for food.

To bring food back to the nest, the Swift gathers insects in a ball in its throat (known as a bolus). If you see a Swift up close, you can sometimes see this bulging food-pouch – like a pocket stuffed full of goodies.

When weather makes insects scarce, parents may have to leave the brood for up to 48 hours to find food. Young Swifts have a unique way of surviving these times, going into a short-term hibernation called 'torpor'. Their temperature falls, and their pulse and breathing slow. Their growth slows right down, reducing their energy needs, so they can survive on fat reserves. Swifts can even go into torpor inside their eggs, if parents are away for a long time.

Once they're around four weeks old, Swift nestlings exercise their wings by doing press-ups on their wingtips. At first, when they press down on the floor, they can't lift their bodies clear of the ground. But soon they can do a 10-second press-up – their wings are strong enough to fly.

Around a week before they leave the nest, young birds spend a lot of time at the entrance, gazing out into the wide world, spotting passing insects and feeling the wind. In his portrait of the Swift, *On Crescent Wings*, wildlife artist Jonathan Pomroy describes watching nestlings rush back into safety at any sudden loud bang, such as a van door slamming shut. Gradually the young grow bolder, finally sticking their heads out, sensing when the time is right to leap. Parents leave them for a few days, so their hunger spurs them on. Nestlings often leave the nest at twilight – as a screaming party of 'teenagers' whizzes past, they leap, freefalling from the nest and opening their wings to flap furiously into the heights to join them.

The young leave the nest once and for all. Within days of fledging, they'll be flying to Africa. Only in their third or fourth summer will they find a nest-site to breed. In all that time, they will not once have touched land.

A FLYING VISIT

Swifts are the last of our summer visitors to arrive and the first to leave, as soon as their young have fledged. Though Swifts are British by birth, they spend only a third of their lives in Britain. By the end of August, all Swifts will have left Britain, and none will be seen again until the following April or May.

SWIFT'S LONG JOURNEY

A Swift migrates hundreds of miles a day on its 6,000-mile journey to Africa (one recorded Swift flew 620 miles in a day). The journey takes the bird through Europe and beyond the Sahara (where Swifts pull a special membrane down over their eyes to protect them from sand).

Once in Africa, the Swift flies over savannah and rainforest, where hippos wallow in wetlands, chimpanzees nest in trees and elephants tramp the land. Even then, the Swift doesn't spend winter in one place, but travels around, chasing food and racing weather. Over a lifetime, a Swift may fly a distance equivalent to going to the moon and back five times!

HELPING SWIFTS – FIT A SWIFT BOX

Between 1995 and 2015, there was a 15 per cent drop in Swift numbers, putting them on the 'red list' of Birds of Conservation Concern. Monoculture farming and pesticide use has meant a loss of insect-rich habitats and a build-up of poisons in the birds' bodies. The demolition and modernisation of old buildings has had a huge effect on Swift numbers, leaving the birds without the eaves, crevices and cavities they need to nest in.

After retiring, Swift-lover John Stimpson began making wooden nest boxes in his garage. Thirteen years later, he reached his goal of making 30,000 Swift boxes, enough to house half the United Kingdom's breeding population. You can buy one of John's boxes from the organisation Swift Conservation (page 189).

RECLAIM THE SWIFT – RENAME THE SWIFT!

The Swift's cry has commonly been called a 'scream'. It is loud, and not very musical, yet I like to think of it as a squeal of delight – like someone riding a rollercoaster. Some of the Swift's old names were downright derogatory, and reflected people's own religious beliefs, rather than the reality of the bird. So, I invite you to reclaim the Swift's names. Invent your own names, inspired by the birds. Children can create Anglo-Saxon-style kennings, which join two words to create a vivid new name (like a kind of mini-riddle), such as 'sea-rider' for ship, or 'sky-candle' for sun. Celebrate the Swift with words!

SWIFT – SOME NEW NAMES

Rain-racer

Scythe of the Sky

Black Bow

HOW TO DISTINGUISH BETWEEN SWALLOWS, HOUSE MARTINS AND SWIFTS

To spot the difference between these three close cousins, look carefully at each bird's shape and colour, and the way they fly.

Generally, Swifts fly the highest of the three birds. House Martins often flutter around at rooftop height. Swallows often swoop low and dart in different directions.

A little ditty to help you remember:

*Swifts fly high
Swallows swoop low,
And dart to and fro.*

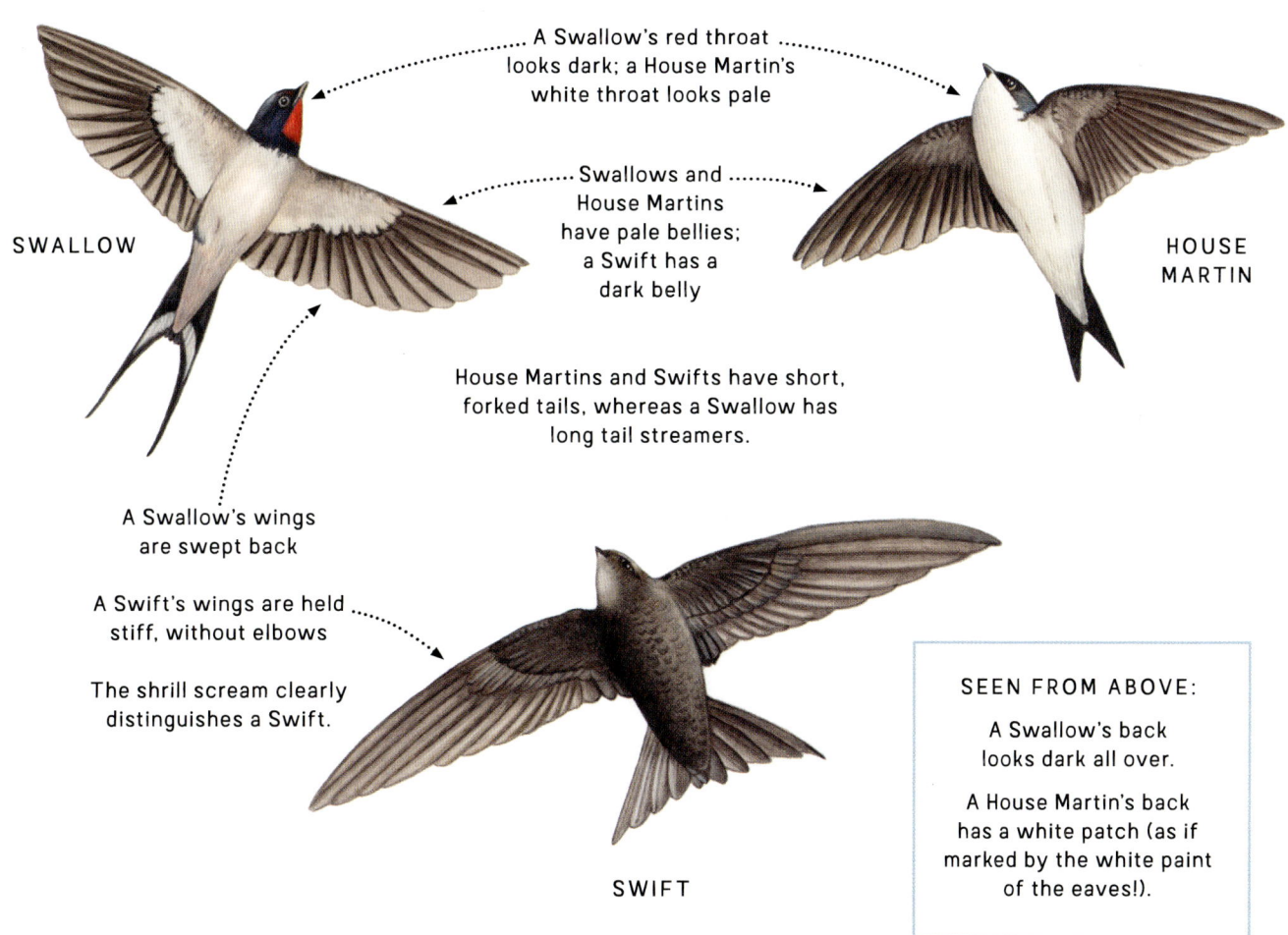

SWALLOW

A Swallow's red throat looks dark; a House Martin's white throat looks pale

Swallows and House Martins have pale bellies; a Swift has a dark belly

HOUSE MARTIN

House Martins and Swifts have short, forked tails, whereas a Swallow has long tail streamers.

A Swallow's wings are swept back

A Swift's wings are held stiff, without elbows

The shrill scream clearly distinguishes a Swift.

SWIFT

SEEN FROM ABOVE:

A Swallow's back looks dark all over.

A House Martin's back has a white patch (as if marked by the white paint of the eaves!).

It is quieter now. In the heat of the afternoon, only the
pigeon coos. Already, things are looking a little past their peak,
in the baking heat. The flowers have turned into fruits. On the
branches, there are small round apples. On the brambles, the berries are
green and red, and – perhaps – even turning black. The scent of privet
is heavy in the hot, still air. Sometimes a summer heatwave is broken
by a summer storm – nice weather for ducks!

In this chapter, we celebrate our most familiar wild waterbirds,
the Mallard Duck and the Mute Swan.

JULY

MALLARD

Anas platyrhynchos

MALE

Green head

Yellow bill

Black-and-white tail – the middle feathers are curly

White collar

Shiny purple-blue wing patch

Brown breast

FEMALE

Dark eye-stripe

Shiny purple-blue wing patch

Patterned brown plumage

Dull bill, with patches of orange

When moulting (July–September), the male looks similar to the female – recognise him by his yellow bill.

VOICE

The familiar *quack-quack* of the Mallard is made only by the female. Males make a quieter nasal call. Ducklings *peep-peep-peep* as they follow behind their mother.

SIZE

50–65cm

WHERE

On/near water.

NEST

Female, lying on belly, digs with feet to make nest on ground. Adds leaves, grasses and down plucked from her own breast.

EATS

Water plants, cereals, insects, fish.

MALLARD– THE OLD NAMES

Common Duck – the most common duck on Earth

Wild Duck – the Mallard is the wild ancestor of almost all farmyard ducks. Wild ducks were once hunted, and domestic ducks farmed for food (since Neolithic times)

Mire Duck – meaning marsh duck (Forfar)

From Proto-Germanic *musą*, 'bog/moss': **Moss Duck**

Dabbling Duck/Puddle Duck – since a Mallard feeds by dabbling

COMMON DUCK

The much-loved Mallard is our most familiar duck, at home on park ponds, city canals and country rivers, as well as remote reservoirs. For many young children, feeding the ducks is the first direct contact they will have with wildlife. Mallards are often tame, coming close to humans and even taking food from the hand.

Mallards are 'dabbling ducks' – they sift tiny plants and insects from the surface of the water with their broad, flat bills. Along the inside of the bill, little combs (that look like tiny teeth) sieve out water and mud, and keep in food, like seeds and bugs. In deeper water, watch Mallards upending – the duck equivalent of a headstand. With heads underwater and tails in the air, they stretch their necks to reach the food below. As Ratty sings, in *The Wind in the Willows*:

> *All along the backwater,*
> *Through the rushes tall,*
> *Ducks are a-dabbling,*
> *Up tails all!*

You can tell a Mallard drake (male), even upside down – its tail is black with white edges, and the central feathers curl up at the tip.

DUCKS AND DRAKES

Male Mallards are known as drakes, females as ducks. 'Ducks and Drakes' is the old English name for a simple waterside game:

- Find a flat stone.
- Skim it across the surface of the water.
- How many times does your stone bounce? The world record is 40!

Just like real ducks and drakes, thrown stones skim the water, making rippling rings and splash landings.

LOVE A DUCK!

During courtship, several drakes gather around a duck, and woo her with elaborate displays, with delightful names like grunt-whistling and head-up-tail-upping. Look for courting males in October and November. You might see:

- **Bill dipping and water flicking**: A male flicking the tip of his bill through the water, sending up an arc of spray.
- **Head-up-tail-upping**: A male lifting both wings and tail, with a loud whistle.
- **Mock-preening**: A male lifting one wing, showing off his bright wing patch, and running his bill along the quills, making a loud rattling noise.

If a courting drake has attracted a duck to pair up with him, you might even see the drake leading away from the crowd, with the duck following after.

MAMA DUCK, FLUFFY DUCKLINGS

Fluffy, yellow ducklings, peeping loudly as they follow their mother, are a delightful sight. The nursery counting rhyme 'Three Little Ducks' celebrates ducklings:

Three little ducks went out one day,
Over the hills and far away.
Mamma duck said
Quack-quack, quack-quack.
But only two little ducks came back.

Two little ducks went out one day
Over the hills and far away.
Mamma duck said
Quack-quack, quack-quack.
But only one little duck came back.

One little duck went out one day,
Over the hills and far away.
Mamma duck said
Quack-quack, quack-quack.
And three little ducks came swimming back.

Spot ducklings in spring and summer. The mother does not bring food for her newly hatched ducklings. Instead, after just one day, she leads them from the nest to water, where they can forage for themselves. Ducklings can swim the day they hatch and know instinctively how to escape danger from above by diving underwater. Their pale underbellies camouflage them from predatory fish like pike. If chicks are threatened by a ground predator, such as a fox, the mother drags her wing, pretending to be injured, to distract the fox's attention from her young. Ducklings stay with their mother for up to two months, until they're able to take care of themselves.

The mother duck is a symbol of the divine mother in both ancient times (page 117) and recent times – in Scotland, it was taboo to hunt a duck swimming with her young, as such a duck was sacred to Mother Mary.

WATER OFF A DUCK'S BACK

In spring, the drake's dashing feathers help attract a mate. Feathers must also be waterproof, so, to keep them in top condition, ducks grow a new set each year, shedding their old feathers between June and September. Summer is the ideal time to look for feathers by the waterside.

During the annual moult, drakes lose their colourful plumage and instead wear 'eclipse' plumage. Until their new flight feathers grow, they're unable to fly, and the patterned browns of the eclipse plumage – which are similar to the female's feathers – camouflage them from predators.

From troubles of the world
I turn to ducks,
Beautiful comical things ...
FROM 'DUCKS', F. W. HARVEY

NICE WEATHER FOR DUCKS

As waterbirds, ducks don't mind the rain, and feeding the ducks is a good rainy-day outing. In Ireland, many of the duck's behaviours, including waddling, quacking, flapping and flying, were all said to forecast rain.

A folktale from Mayo tells how the duck got its waterproof feathers: Once, Jesus was passing by where a duck and a hen were feeding. There was a sudden heavy downpour. When Jesus asked for shelter under the hen's wing, the hen pecked Him. But the duck allowed Him to shelter under its wing until the storm had passed. In gratitude, Jesus made the duck's feathers waterproof, and they've been waterproof ever since. A similar tale from Uist tells how the duck got webbed feet. Jesus was once being pursued, and a crofter helped Him hide in a heap of corn. The farmyard ducks and hens came flocking to eat the grain. When the ducks ate, their feet trampled down the grain, keeping Jesus hidden. But when the hens ate, their feet scattered the grain far and wide, so the hiding place was revealed. Because of this, ever since, the duck has been blessed with webbed feet, which give 'three joys': 'the joy of earth, the joy of air and the joy of water'. According to the story, the blessing of 'the joy of water' gave the duck a love of 'hail and rain [and] thunder and lightning'. It was said of the duck that 'when she hears thunder she rejoices and dances to her own *puirt-à-beul*' ('mouth music' – a lilting, cheerful tune sung to accompany dancing, in the absence of musical instruments), hence the saying: 'Thou art like a duck, expectant of thunder.'

A dance known as 'the waddling of the ducks' was once popular in the Highlands of Scotland – on a rainy welly-wearing day, children might enjoy making up their own duck dance, especially if the 'dance floor' includes plenty of puddles!

SACRED WATERBIRD

The way that ducks can both fly and swim, uniting 'the joy of air and the joy of water', may be why, from the Bronze Age to the Celtic Iron Age in Europe, ducks (as well as other waterbirds) sometimes appear alongside symbols of the sun. One image often used on Bronze Age metalwork, for example, was a boat shaped like a duck, carrying the disc of the sun.

I wonder if the first boats were inspired by ducks, and their ability to float on the surface of the water rather than sink below it. The keel of a boat is shaped like the breastbone of a duck (which is also called a keel). In Celtic Gaul, ducks were associated with the goddess of the river Seine, Sequana. A bronze statue of Sequana was found at her shrine, at the river-source where sacred springs were revered for healing and restoring health. She stands on a boat shaped like a duck, with a duck-head prow and curly-tailed stern.

Perhaps an echo of these old ideas still rings in the Scottish belief that it is lucky to see a wild duck on New Year's Day, especially for a sailor – it means he will not drown.

FEED THE DUCKS

Bread is not good for ducks – it fills them up and so stops them from foraging for the nutritious foods they need. Instead, feed the ducks with:

- Sweetcorn.
- Lettuce.
- Oats.
- Seeds.
- Peas.

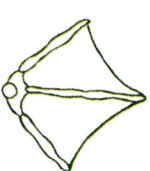
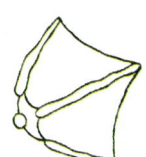
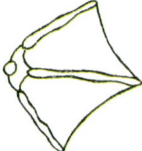
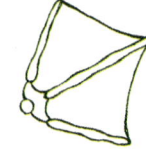
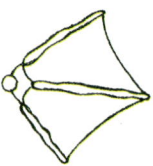

'MAMA DUCK HATCHES THE WORLD' A CREATION MYTH FROM THE FINNISH EPIC POEM, THE KALEVALA

A story that honours the mama-magic of the female Duck

In the beginning, there was only water.

Nothing but water.

Above the water, flew a duck. A little brown duck, looking for somewhere to land, somewhere safe to make her nest and lay her eggs. She flew east and she flew west. Nothing but water. She flew north and she flew south. Nothing but water.

In the water, the great goddess, the Mother of the Sea, was watching the duck. She felt for the little brown bird. And so, she lifted up her knees, out of the water, to make a place for the duck to land. Down flew the duck. She tucked in her wings and landed on the Sea Mother's lap. There, the duck made her nest. She ruffled her feathers and settled there, content. The Mother of the Sea was happy too, to have a feathered friend. She stroked the duck's feathers and talked softly to her.

And the duck laid seven eggs; six eggs of gold, and the seventh egg of iron. The Mother of the Sea had never seen an egg before. She marvelled at them. Very gently, she touched them. They seemed lifeless; hard and cold.

But the mother duck snuggled her eggs to warm them. The eggs grew warmer and warmer. The nest grew warmer and warmer. The Sea Mother's legs grew warmer and warmer, hotter and hotter. Until she felt her skin was scorching, and she shifted her legs. But the nest tipped ...

The duck flew, up, into the air, and the eggs fell, down, into the water. Down ... down ... down ... *CRACK*! The seven shells, bright gold and rich iron, broke open.

But in that moment was a wonder. The broken pieces came together, making two great egg shell pieces; one the upper, one the lower. One great egg, broken open.

With eyes wide, the duck stared! And still the wonder-change continued. The lower part of the egg became the earth beneath our feet. The upper part of the egg became the sky above our heads. The yellow of the yolk became the gold of the sun. The white of the egg became the silver of the moon. The speckles of the shell became the sparkles of the stars. And so the world was born. And how that mama duck puffed up her feathers with pride!

The Mother of the Sea watched the duck and she smiled. She looked out at the new world and lifted her face, feeling the warmth of the new sun.

With her companion the duck flying beside her, the Mother of the Sea explored the shore. Where she trod, her feet made the caves and the caverns. With her hands, she shaped the lands; she made the coasts and the curving bays.

That is how the world began, hatched from the egg of a little brown duck.

MALLARD

MUTE SWAN

Cygnus olor

Long curved neck

White feathers

The male
Mute Swan
has a bigger
bill-bump than
the female ....

Orange bill,
with black
bump at base

VOICE

Mute Swans are not really silent.
Though they don't sing, you may
hear them make soft contact calls,
or grunt or hiss if feeling defensive.
In flight, a Mute Swan's powerful
wingbeats make a loud throbbing
hum, which acts as a contact call.

SIZE

125–155cm
– one of our
biggest birds
and one of the
world's largest
flying birds

WHERE

On water.

NEST

Mound near
water.

EATS

Water plants,
grasses, insects,
snails.

MUTE SWAN – THE OLD NAMES

From the sound of the Mute Swan's wings in flight:

Swan – a name unchanged for over 1,000 years – related to Sanskrit *svana* and Latin *sōnus*, 'sound'

Tame Swan – As the Mute Swan was once domesticated, it was given the name Tame Swan, to distinguish the bird from the winter visitor, the Whooper, which was known as the Wild Swan

Mute Swan – Another name given to distinguish the bird from the Whooper Swan, which has a louder call

THE WHITE BIRD OF THE WATERS

See Mute Swans on ponds, lakes, reservoirs and slow-moving rivers and canals, all year round. Railway lines often follow the course of a river through a valley – on train journeys children can watch from the window for Mute Swans.

Mute Swans nest close to the water, amidst the cover of riverside plants. The cob (male) brings reeds and rushes, sticks and twigs, and the pen (female) builds a huge mound – up to 4m across. When the pen is sitting on eggs, the cob is a fierce protector – raising his wings and hissing to deter intruders.

LOVEBIRDS

Mute Swans are famously loyal couples – only 3 per cent of breeding pairs separate. Young swans spend at least two years with the flock before they pair up. Even then, they have a long 'engagement', waiting another year before they breed. Mute Swans reaffirm their bonds at the beginning of every breeding season, with courtship displays throughout spring and early summer. One courting gesture is known as 'head-turning' – two swans face each other, heart to heart. Their arched necks make the shape of a love heart.

SWAN RIDE

A cygnet's fluffy down is grey-brown. The first day of its life is spent on the nest with its parents. Cygnets mostly sleep, sheltered under the mother's tummy or wings, occasionally tottering around nearby, exploring their new world. You may be lucky enough to see a whole family of Mute Swans, the mother leading the way, cygnets following after, and father keeping guard at the back.

Anyone who has taken a toddler for a walk knows that there comes a point when little legs start to feel tired. The water can feel especially cold to cygnets, whose fluffy down is still thin. If they've been swimming for a long time, they often seek the shelter of a parent's back. One parent (usually the pen) lowers its tail so the cygnets can climb aboard. Sitting between her wings, the cygnets are warm and safe, and if they're really tired, can even snuggle down and have a nap. In heavy rain, the pen positions her wings like an umbrella to shelter them. The cygnets can even stay in place when the mother upends, holding on with their beaks as she reaches her long neck underwater to find food.

Swans sleep on the water, where there are fewer predators. At night, cygnets often sleep upon a parent's back, protected from danger and from the cold.

SWANNING AROUND

The Mute Swan family roams through its territory. The female leads the way to spots where she knows there's plenty to eat. Both parents find food for their young, pulling up underwater plants with their long necks and laying them on the surface where the cygnets can reach them. In shallow water, you might see adult swans 'foot-trampling'. The adults paddle on the spot so powerfully that they rise up out of the water. Their huge webbed feet stir up the lake or riverbed, so that insects and pieces of plants float to the surface for the young to eat. Sometimes the pen uses foot-trampling to call her hungry babies, like a mother calling 'supper-time!'

JOINING THE FLOCK

Cygnets stay with their mother and father for 4–5 months, before leaving their parents' territory and joining a flock. Sometimes youngsters leave in a group, two or three siblings travelling together. Flocks are made up of young (non-breeding) birds along with adults without territories.

Youngsters who have just left their parents spend two or three years in these flocks, learning all about the life of an adult Mute Swan. In a large flock, you may spot a few pairs sticking close to one

another – these are older members who have paired up, but have not yet left to look for a territory of their own.

Historically, all three species of swans (Mute, Bewick's and Whooper) gathered together at dusk to roost – a spectacular sight (and brilliant sound – the 'bell-beat of their wings') celebrated by Yeats:

> *The trees are in their autumn beauty,*
> *The woodland paths are dry,*
> *Under the October twilight the water*
> *Mirrors a still sky;*
> *Upon the brimming water among the stones*
> *Are nine-and-fifty swans.*
>
> FROM 'THE WILD SWANS AT COOLE'

BIRD OF SPIRIT, BIRD OF MYTH

Images of swans appear on pottery and metal-ware from the Bronze and Iron Ages, and even in Stone-Age rock carvings. (The oldest cave art in Britain, carvings on the Creswell Crags in Derbyshire, include a swan-like figure.) With its grace and grandeur, plumage of purest white and singing wings, the swan has always made a deep impression on the human imagination.

In Sanskrit texts of ancient India, and ancient Hindu and Buddhist art, the *'hamsa'*, a white water-bird (which many scholars interpret as a swan), is associated with the pure light of the divine-spirit – 'the highest soul … that pervades the universe and shines like 10 million suns'.

In Norse mythology, the water of the sacred well in Asgard, home of the gods, is so pure that everything it touches turns white. Two swans drank from the well, and that is why swans have been pure white ever since.

In old stories, the idea of the swan-maiden – a woman who can take the form of swan – is very ancient and very widespread, across Europe and beyond. Ireland has a particularly rich hoard of tales of magical swans – humans transformed into birds; you can tell enchanted swans by the golden chains or crowns they wear. Later tales drew on these old legends. Hans Christian Andersen's *The Wild Swans* stems from an Irish folktale – it tells of 12 brothers cursed to become swans, freed by their sister. The Brothers Grimm also have a version, *The Six Swans*. These stories have inspired many other forms of art, such as the ballet *Swan Lake*.

BIRD OF GOOD OMEN

In Alexander Carmichael's book of folklore of the Highlands and Islands of Scotland, *Carmina Gadelica*, he shares that a swan is a bird of good omen: 'To see seven, or a multiple of seven, swans on the wing ensures peace and prosperity for seven, or a multiple of seven years.' To see a swan on a Friday morning, at sunrise, is an especially good sign:

> *If you see a swan on Friday morn,*
> *As night turns into day,*
> *You know that you will prosper*
> *Throughout the year to come.*

THE GODDESS AND THE BARD

Brigit is the Celtic goddess of the sacred flame and the holy waters. In Ireland today, she is celebrated as both goddess and saint, and often depicted wearing white (an Irish rhyme sung at the festival of St Brigit tells, 'Here comes Bridget dressed in white'). The goddess Brigit is associated with the swan – the white bird of the waters – by Robert Graves and by modern pagans.

The ninth-century encyclopaedia of Irish oral tradition, *Cormac's Glossary*, tells us that the cloaks of poets were made of 'the skins of birds, white and many coloured'. The swan-feather cloak of the bard is a vivid image of the mystical magic of story and song.

ROYAL BIRD

In England, the swan has been a royal bird since ancient times – a delicacy for royal feasts. For centuries the Crown claimed ownership of all swans on the River Thames. Royal swanherds pinioned the birds (removed their flight feathers) to stop them from flying away. No subject was allowed to own swans on a public river, except by a grant from the Crown. To bestow this privilege, the Crown granted a swan-mark – a mark nicked onto a swan's bill as a sign of ownership.

Two of London's oldest livery companies, the Vintners' Company and the Dyers' Company, were granted their own swan-marks. The Vintners' swan-mark was a 'V' formed of two nicks – possibly the origin of the pub name the Swan with Two Necks ('Nicks'). Other centuries-old British inns include the Swan, the White Swan and the Swan Inn.

By tradition, swans' bills were marked in a ceremony known as swan-upping. Young swans were rounded up and lifted up out of the water (hence swan-*upping*) to nick their bills with a knife.

Nowadays, owning and marking swans for food is a thing of the past. But the Vintners' and Dyers' Companies continue the custom of catching swans on the Thames – they monitor numbers for conservation and remove any harmful fishing hooks or lines, releasing the swans back onto the river unharmed.

SOUL-BIRD

In Irish folklore the spirits of the departed were sometimes said to visit loved ones in the form of a bird. To injure or kill a swan, a soul-bird, was believed to bring great bad fortune. Still today, for modern Druids, the swan is a symbol of the soul.

'THE HEALING SWAN' A FOLKTALE FROM IRELAND

A story celebrating the swan

In the high blue hills lay a green glen. Sunk in the slopes was a low stone home, with its back to the wind and its face to the sun. The floor was earth and the roof was round, thatched with reeds.

Early one morning, the mother of the house went out. Down the track to the loch, to fetch a pail of water, to make broth for her son.

Her little boy was not well. He lay in his bed, too sick to sit up, or to eat or even to smile. He slept fitfully, twisting and turning, tangling the bedclothes. His face was thin. His cheeks were pale. His eyes were dull, smudged with dark shadows.

The mother sorrowed, but she kept the peat fire burning bright, to keep him warm. She gave him hot broth and healing herbs. She prayed with all her heart. But the boy had not got any better.

As the child slept, his mother trudged through the creaking snow and the stinging wind.

Upon the white snow, a stain of red … A shout of colour. There were splashes and spatters of red – a trail of blood, leading towards the half-frozen lake. The woman put down her pail, and she followed the trail.

In the summer, the reed bed was gold and green, whispering and waving in the breeze. Now, it was grey and silent.

There, curled up in the marshy shallows, she found …

… a swan. Hurt. Its wing was wrong, crooked. Its foot was cut, pierced by a sharp fishhook. Its neck was tangled in the fishing line.

The woman looked at the wounded bird. 'Dear thing.' Very slowly, she knelt. Very gently, she unwound the line from the swan's neck. Very carefully, she took the hook from the swan's foot.

'Come …' She lifted the bird in her arms. He was limp, heavy. She carried him home.

Back home, her son was stirring. He murmured in his sleep and cried out. The woman put her cool hand to his forehead and he closed his eyes.

The mother stroked the swan's curved head, his curled feathers.

She fetched water and she washed the swan's bloodstained feet.

She bathed and bandaged the swan's broken wing.

She filled a tub with cold clear water and golden linseed. And the bird lowered its bill, nibbling, scattering sparkles. The woman laughed. The boy smiled in his sleep.

Day after day, the mother tended the swan.

On the first day, the woman made a good fire of the best peats. She made a bed of blankets and the swan lay, thawing, in the firelight. In his bed, the boy felt cold and he felt hot, he shivered and sweated.

On the second day, the swan ate and ate until all the seed was gone. And the boy opened his lips, and drank broth.

On the third day, the swan folded his wings and tucked his beak into his feathers. He slept for a long time. The boy slept too, deep and sound.

That evening, at the same moment, the boy and the swan opened their eyes. The swan shook out his wings and sensed the air. He paddled across the flagstones. The woman opened the door. The swan walked out into the moonlight.

The swan's feathers were no longer stained with dark blood. They were white as the snow of the hill. The swan shone in the moonlight.

And the boy sat up in bed. He yawned and he stretched. His eyes were bright. He threw off his bedclothes and got to his feet. He was well again. Barefoot, he padded across the flat flagstones. He ran to his mother, and she opened her arms to receive him.

The woman and the boy stood together in the doorway, watching the swan.

They watched the swan walking into the waters of the lake. The wind breathed and the swan sailed, resting on his own reflection, trailing silver ripples.

The woman watched the swan and she began to hum. She made a song, a moonlight lullaby, to bless the swan, and to thank him:

Oh, White Swan,
Hoo hi, ho oh.
Fair White Swan,
Hoo hi, ho oh.
Life and health,
Hoo hi, ho oh.
Peace and Joy,
Hoo hi, ho oh.
Go with love,
Hoo hi, ho oh.
Go with love,
Hoo hi, ho oh.

As the woman sang, the swan stretched out his wings. His feathers beat the wind, his feet beat the waves, he went skimming along the surface of the water, rising.

His flying was his singing – the rhythm of his wings humming, thrumming.

He flew through the night, white in the darkness, shining.

Until all that was left was the dark water, rippling in the moonlight.

It is high, dry summertime. In the heat of the day, the trees are heavy, the leaves dark. In fields and on verges, grasses and grains are golden with seed. In trees and hedgerows, nuts bake brown in the heat, and berries ripen – the year's first blackberries may already be sweet.

These are the dog days of summer – nature is winding down. In garden, woodland and field, the birds are quiet now. For most birds, the work of defending their territories and raising their young is done, and there is time to let go of old, worn-out feathers.

This chapter explores the wonder that is a feather. Walk in the cool of the woods, and perhaps you'll find a feather from the Great Spotted Woodpecker, whom we also meet in this chapter.

AUGUST

Feathers

Birds are the only creatures on Earth that have feathers. Feathers are made of keratin, the same thing our own hair and nails are made of. Feathers enable birds to fly, keep them warm and dry in the cold and wet, and protect them from the heat of the sun.

Feathers can be colourful, to attract a mate, or announce to other birds, 'I'm here. This is where I live.' They can be dull, or speckled, to provide camouflage. A bird's 'down' feathers (those closest to its body) are so soft and warm that for hundreds of years people used them to stuff pillows and quilts – quilts were known as 'eiderdowns', because the softest and warmest down feathers are from Eider ducks, seabirds that breed in the Arctic. The feathers were gathered from the birds' nests, after the ducklings had left.

The Summer Moult

To keep their plumage in good condition, most birds grow a new set of feathers each year. The old feathers are shed by 'moulting'. Most birds shed their feathers a few at a time, over a number of weeks, so that they always have enough feathers to fly. As their new feathers are growing, birds shelter quietly in the bushes, so predators can't see or hear them. Many birds, such as House Sparrows, moult into duller coats. This helps them stay camouflaged and means they save their 'best dress' for the breeding season, so that when they're courting, their new coats are smart and bright.

In spring, birds were busy finding a place to live and a mate to partner with. By summer, eggs have been laid and young hatched. Food is plentiful. The challenges of winter are far ahead. The high dry days of summer are the perfect time for moulting. The school summer holidays, in August, are a good time to spot feathers. The guide on pages 128–129 can help you identify feathers you might come across when you're out and about. At the time of writing, the RSPB advises against touching or collecting feathers, to minimise the risk of spreading avian flu.

Where to Look for Feathers

- Look for feathers on the ground, near trees or hedges where birds perch. Sometimes the feather shaft sticks into the earth when a feather lands, so you can easily spot the feather sticking upright, like a flag.
- In woodlands, feathers often catch in the undergrowth, where they're less likely to be blown away in the wind.
- Look on the shores of ponds and lakes to find feathers of ducks, geese and swans.

Feather Clues – Different Types of Feathers

A bird has many different kinds of feathers:

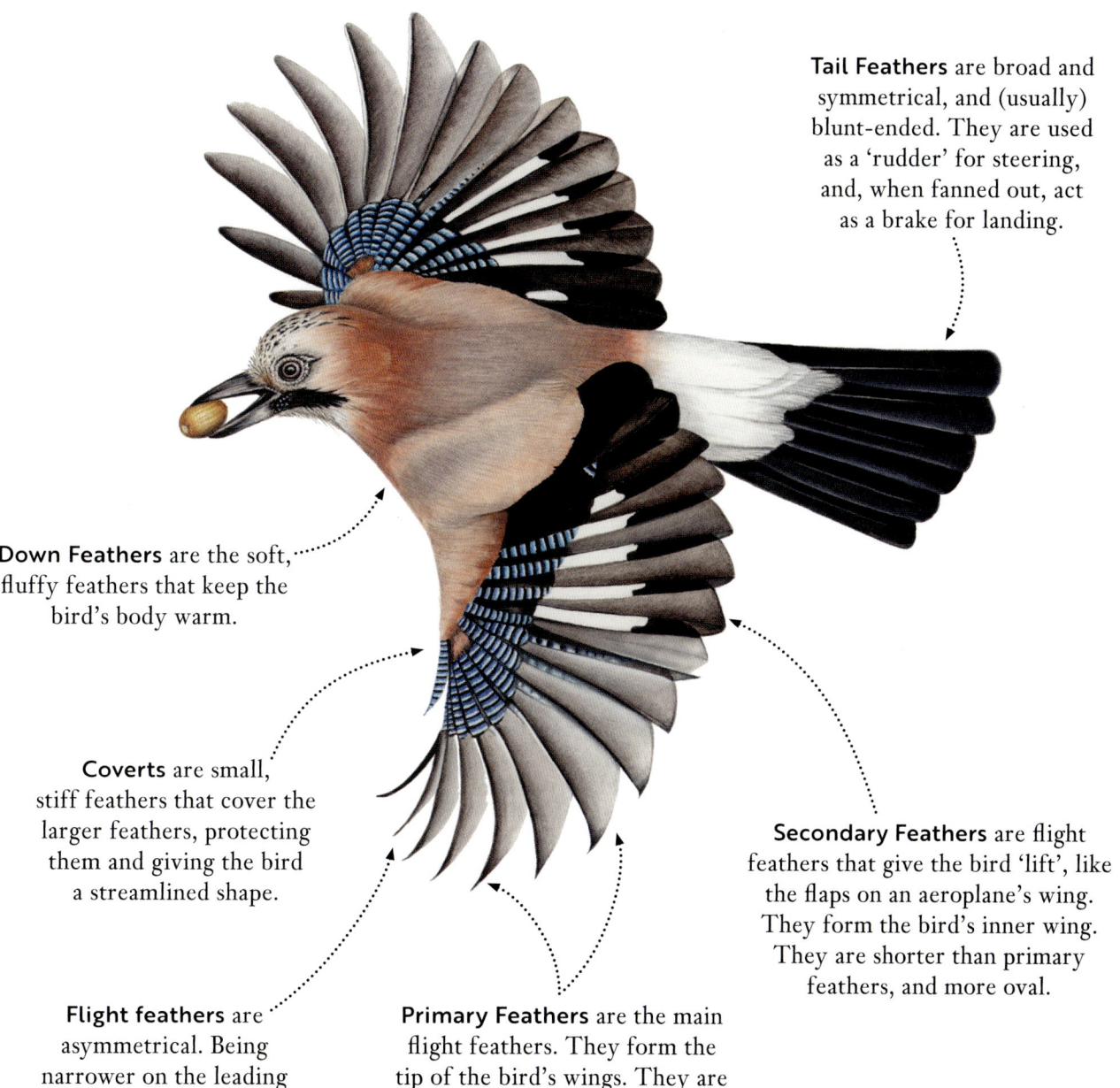

Tail Feathers are broad and symmetrical, and (usually) blunt-ended. They are used as a 'rudder' for steering, and, when fanned out, act as a brake for landing.

Down Feathers are the soft, fluffy feathers that keep the bird's body warm.

Coverts are small, stiff feathers that cover the larger feathers, protecting them and giving the bird a streamlined shape.

Secondary Feathers are flight feathers that give the bird 'lift', like the flaps on an aeroplane's wing. They form the bird's inner wing. They are shorter than primary feathers, and more oval.

Flight feathers are asymmetrical. Being narrower on the leading edge helps the feather slice through the air.

Primary Feathers are the main flight feathers. They form the tip of the bird's wings. They are long and strong, to power the bird's flight.

Feather Clues – Which Bird?

The colour, pattern and shape of a feather all give us clues about which bird it comes from.
Common finds include:

Great Spotted Woodpecker (Primary and Secondary Feathers)

White spots along edges

Black background

Mallard Drake (Secondary Feather)

Glossy purple-blue

Black bar

White tip

Brown-grey

Magpie (Tail Feather)

Top side glossy green-black

Purple-blue band at tip

Black underside

Tawny Owl (Secondary Feather)

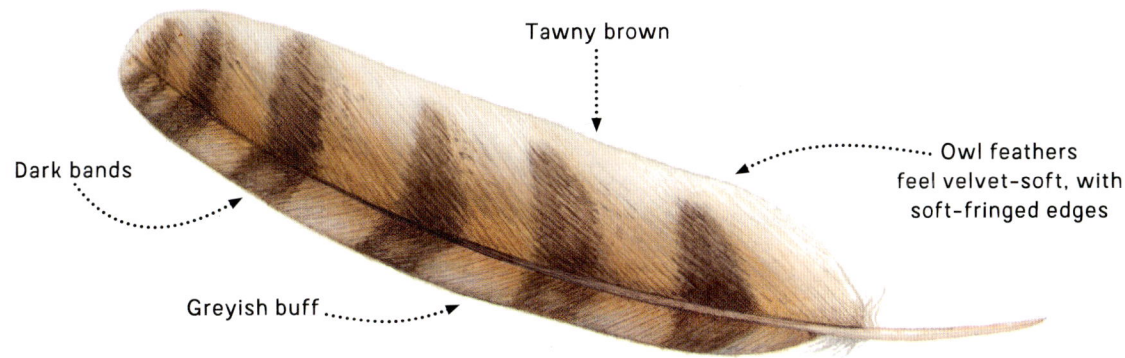

Tawny brown

Owl feathers
feel velvet-soft, with
soft-fringed edges

Dark bands

Greyish buff

Woodpigeon (Tail Feather)

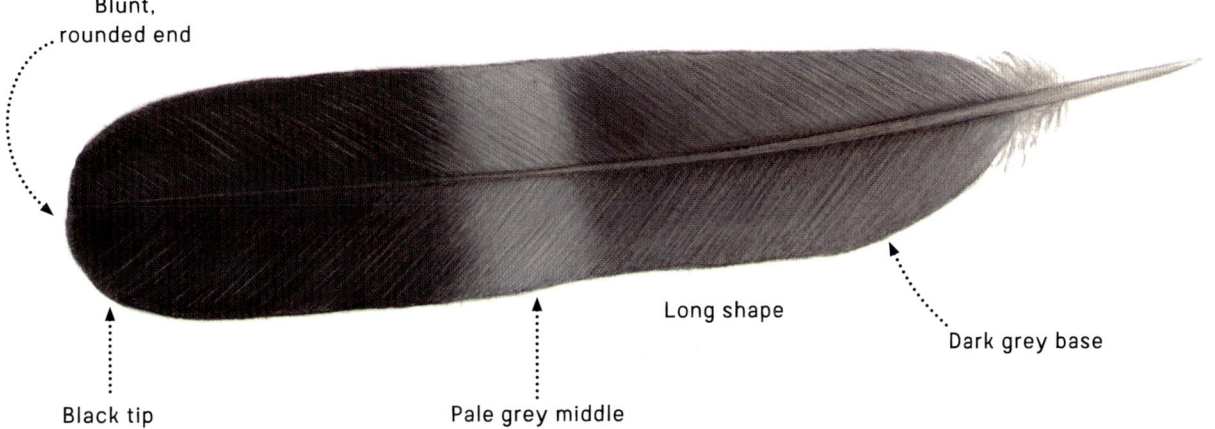

Blunt,
rounded end

Long shape

Dark grey base

Black tip

Pale grey middle

Jay (Covert Feather)

Patch of bright blue with black bars

Feather Clues – Moulted or Hunted?

The feathers we find tell a story. Did the bird shed its feather during moulting? Or was the bird food for a fox? Or for a bird of prey? Looking carefully at the feathers you find can give you some clues.

Moulted feathers often look old and worn. The colour may be faded. The end of the shaft may be dry and cracked.

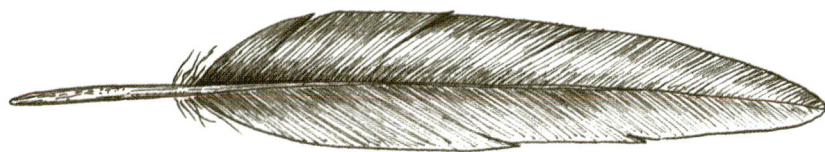

Moulted feather

Before eating their catch, birds of prey use their beaks to pluck out feathers. Feathers are often undamaged except for a bend or break in the shaft where it was gripped in a sharp beak. You might even find a bird of prey's 'plucking post', with a mass of feathers below.

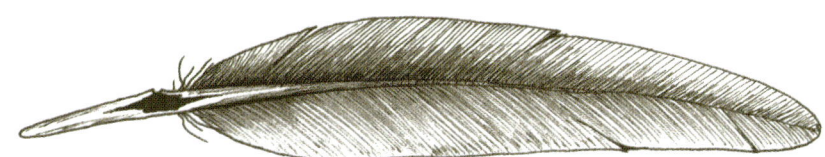

Plucked feather

Foxes, and other mammals that eat birds, cannot pluck out individual feathers. A fox bites whole mouthfuls of feathers at once, and spits them out, often leaving clumps of feathers stuck together with saliva. Feather shafts may be bitten off, or plumes bitten in two.

Feather clump

Pheasant tail feather

FEATHER TREASURES

The feather is a marvel. Feathers are both strong and light. They can be coloured, patterned, iridescent – with tints that shift and change. A feather is a physical link to a creature that has inspired wonder in our hearts since the dawn of time. Throughout history, and around the world, humans have valued feathers. As well as using down for bedding, feathers have been prized for craft, decoration and ceremony.

Arrows The bow and arrow has been used by humans to hunt food since prehistoric times. And the best way to help an arrow fly? A feather, of course. Feathers are attached to the back of the arrow (a process known as 'fletching') to help it fly far and fast, straight and true. An Iron Age arrow recently found in the mountains of Norway had been frozen in ice for over a thousand years – its feather fletching was still perfectly preserved.

Quills The word 'pen' comes from the Latin word *penna*, which means 'feather'. Our earliest pens were made from the moulted feathers of geese and swans. To create a point for writing, the end of the feather was cut at an angle, using a sharp knife (hence our word 'penknife' – a knife used to make a feather pen). A flight feather is both sturdy and supple, and the long hollow shaft is ideal for holding ink. Goose feathers were common finds, and were often used to make quills. A swan-feather pen was especially prized. Crow feathers were sometimes used for quills too – the thin shaft suited fine delicate work, such as map drawing. The words which shaped our world, including the *Book of Kells*, the *Magna Carta*, and the works of Shakespeare, were all written with a quill pen.

Fishing Hooks 'Fly fishing' attracts fish by using bait crafted to look like a fly (or other tempting titbit). A traditional 'fly' is made by fastening feathers onto a fishing hook. With its natural colouring and light weight, a feather fly looks and moves like a real insect on the surface of the water (or beneath the water, carried with the current). Feathers from various birds are used for fly making, including chickens, ducks and pheasants.

Rituals Feathers are symbolic in many spiritual traditions and have been used in sacred rituals across cultures. Throughout history, feathers have helped humans meet the needs of body and soul, keeping us warm and well-fed, and enriching our cultural and spiritual life.

Swan feather

GREAT SPOTTED WOODPECKER

Dendrocopos major

Males have a red patch on the nape

Black back with white shoulder patches

White underparts

Red underneath tail

Females do not have a red nape. Juveniles have red foreheads.

VOICE
The explosive call of the Great Spotted Woodpecker, a loud, sharp *tchick-tchick,* is usually repeated again after a few seconds, and can be heard all year, in places with mature trees. In spring, the Great Spotted Woodpecker's far-carrying drumming is unmistakable.

SIZE
23 cm

WHERE
Near old trees. None in the far north of Scotland. Very few in Ireland, though numbers are increasing.

NEST
Hole in tree (often birch or oak).

EATS
Insects, nuts. At winter bird tables, likes suet, fat, hazelnuts and peanuts.

GREAT SPOTTED WOODPECKER
– THE OLD NAMES

From the Great Spotted Woodpecker's colours:
Black and White Woodpecker (Norfolk) • **Pied Woodpecker (Surrey)** • **Wood Pie (Hampshire)**

From the Great Spotted Woodpecker's pecking:
Wood Hack (Lincolnshire) • **Wood Knacker (Hampshire)** • **Oak Tapper (Ancient Greece)**

DRUMMER BIRD

The Great Spotted Woodpecker's morning drum roll is the soundtrack to the woods in early spring. Parks with old trees are good places to hear woodpeckers too. Though they're usually shy, in springtime you might see courting birds playing an exhilarating game of chase through the canopy.

The male woodpecker is unique among British birds in drumming, instead of singing, to proclaim his territory. The female sometimes drums too, as a contact call or alarm. A woodpecker uses a resonant piece of wood, often a dead, hollow branch or trunk, as a drum, and its beak as a drumstick – one drum roll contains between 8 and 16 pecks per second. The sound is so loud that it can be heard half a mile away. Woodpeckers sometimes drum on telegraph poles, which are, after all, simply long lengths of dead wood. The bird sometimes even plays on a high-tech drum kit, with built-in amplifier – a corrugated metal roof.

WOODLAND CARPENTER

A woodpecker also uses its beak-beating power to find food, using its sharp bill to hack holes in dead and decaying trees to reach the insects and grubs that live there. The beak is as strong and stout as a carpenter's hammer, and has a squared end, like a chisel – a multipurpose tool that can hack a crack into a trunk and prise off flakes of bark with a powerful twist. Just as a carpenter might wear a facemask to protect the face from sawdust, a Great Spotted Woodpecker has special bristle feathers to keep wood-dust out of its nostrils.

A HARD HAT

To reach its food, a woodpecker is essentially banging its head against a tree. The bird's head is well-suited to the job, with a thick skull, and a layer of lightweight, porous bone between beak and brain. This sponge-like bone was once thought to act as a shock-absorber. But scientists have recently discovered that woodpeckers don't seem to need shock absorption after all. The new study shows that the bird's head and beak act like a stiff hammer, helping the bird peck hard and fast with minimal energy.

AN EXTRAORDINARY TONGUE

As well as a stout beak and a strong head, the woodpecker also has an extraordinary tongue. A woodpecker's tongue is so long, it's too big to fit into its own mouth. Instead, a special tube runs through the bird's lower jaw, back beneath its ears and up the back of its head, to the top of the skull. This long 'tongue tube' means the tongue can be rolled away and 'stored' when not in use. A woodpecker's extra-long tongue helps it reach grubs that tunnel deep into trees. The bird pecks open a crack in the wood, then unwinds its tongue to reach right into the tunnel. The tongue is covered in sticky saliva, trapping bugs in a gluey grip. The tip of the tongue is sharply barbed, like the hook at the end of a fishing line – the woodpecker harpoons a grub with the tip of its tongue, then winds it in, to bring its catch to the surface.

GOING UP

A woodpecker's powerful claws are unusual too; it has two toes facing forwards and two facing backwards – see them for yourself with binoculars (page 171). Its even toes give the bird a balanced, powerful grip on upright surfaces like tree trunks (even on the smooth bark of a beech). The woodpecker perches on the side of a tree in an upright stance. Its long, stiff tail acts as a sturdy 'prop', like a walking stick, to help the bird stay stable. The woodpecker moves upwards, vertically or by spiralling up the trunk with jerky hops. Once the bird has finished searching for food on one tree, it swoops to the base of another to begin again.

RAIN BIRD, THUNDER BIRD

The upward-facing posture of woodpeckers has inspired much folklore. In France it was said that the Green Woodpecker was always looking up at the clouds, calling for rain with its cry, *pluie pluie*. A French saying ran *quand le pivert crie, il annonce la pluie* – 'when the woodpecker cries, he announces the rain'. Sound does travel further in cold air – a fact that could explain why people might hear woodpeckers calling more before rain.

The Great Spotted Woodpecker's black-and-white feathers camouflage the bird amidst the dappled shade of trunks and branches. The bird's red patches, though, are bright, brilliant crimson. Birds with red markings have often been associated with fire. Some think of the woodpecker as a bringer of lightning (the fire of the stormy skies), associated with the Norse god of thunder, Thor. Certainly, the woodpecker's resounding drum roll evokes the drumming of rain or the rolling of thunder. The association with thunder could also be because woodpeckers are fond of oak trees – the tree of Thor and often the tree most likely to be struck by lightning.

WOODPECKER'S WOOD NEST

A pair of woodpeckers also uses their sharp bills to hollow out a nest-hole. After a week or two of pecking at the tree trunk, the birds have created a cavity, and work inside the hole to form a pear-shaped nest. A scatter of woodchips beneath an old tree is a clue that woodpeckers have nested above. Look high up on the trunk, or on a stout branch, for a circular hole about 6cm across. From this doorway, a short, horizontal tunnel leads into the nest. After the young have flown, other woodpeckers may use the hole for roosting.

Many other birds rely on holes in trees to nest in. By building a new nook each year, the woodpecker leaves behind its ready-made nests for others to use – a true master carpenter, building safe sanctuaries for many forest folk.

FOREST SIGNS – SPOT THE CLUES THE WOODPECKER LEAVES BEHIND

If you've heard a woodpecker close by, check nearby trees – woodpeckers often hammer near to where they feed. If an old tree has bits of bark missing, look closely; you may see the holes left by a woodpecker's bill.

Look out too for a 'woodpecker workshop'. The birds often use a favourite stump or tree to prepare their food. They wedge a hazelnut into a crack, using the wood like a vice to hold the nut in place whilst they hack it open. You might see the remains of a nutshell (or a pinecone) still wedged into the woodpecker's 'anvil'. Heaps of nutshells, or the remains of cones, at the base of a tree or stump are another clue. When woodpeckers hack open hazelnuts, the shell is shattered – if you find pieces of nutshell with sharp edges, they might be leftovers from a woodpecker's meal. Especially in cold winters in the north, woodpeckers rely on the seeds of pine and spruce trees. They pull the cone apart to reach the kernels within, leaving it looking distinctly dishevelled (or completely demolished) – some of the scales may be split, have broken ends, or be totally missing.

THE NEED FOR OLD TREES

Woodpeckers often peck on wood that is slightly rotten or decaying – it's easier to bore and is a habitat for the insects and larvae that the bird eats. Woodpeckers prefer woods where at least some of the trees are mature, and some are dead or have dying branches. Their nests are often made in trees that have been weakened by fungi. You are much more likely to see woodpeckers in woodland that contains old trees, rather than in commercial woodlands (such as plantations of non-native, fast-growing conifers, planted for timber) where foresters remove old, hollow trees – woodpeckers need decaying wood to survive.

WOODPECKER MAGIC

Pliny and the Brothers Grimm both recorded a belief that the woodpecker is associated with a mythical plant, the 'springwort', which opens any lock (or, in France, it was said, gave superhuman strength when rubbed onto limbs). To enlist the woodpecker's help in finding this mythical herb, the instruction was to plug the bird's nest-hole and place a red cloth on the earth below. When the bird returned to its nest and found it blocked, it would fly off to get the springwort, 'unlock' its door, and then drop the plant onto the red cloth (to return the charm to its magical element – lightning fire). I suggest that anyone who wants to connect with the magic of the Lightning Bird simply look for a woodpecker feather (page 128) – a more humane way to enjoy the wonder of the woodpecker.

TUNING IN TO THE WOODPECKER

Deer can swivel their ears in different directions to pinpoint a noise. The large size and cupped shape of their ears help them pick up the slightest sound that might mean danger. Tune in your own 'deer ears' to help you find woodpeckers in the forest with this simple skill:

- First, cup your hands in front of your ears, with palms facing behind you, then cup them behind your ears, with palms facing forward. Try angling your palms, the way a deer swivels its ears. Notice how each position affects what you hear.
- Sitting by a fire is a good place to practise 'deer ears'. Cup your hands behind your ears and move your head slowly from side to side, like a deer, as you tune in to the sounds of the flames. Try standing facing away from the fire, then turn to face the blaze – notice the difference.
- Whenever you hear something interesting, like a woodpecker drumming, you can use 'deer ears' to amplify your hearing.

'HOW THE WOODPECKER BECAME' A FOLKTALE FROM WALES

A story that helps us to remember Great Spotted Woodpecker's distinctive features – black back, white shoulders and patches of red

Once upon a time there was a little old woman. She was as neat as can be in her smart black dress, with her red flannel petticoat just peeping out beneath. She wore a big white apron, tied over her shoulders with wide white straps. And on her head she wore a smart, red hat. It kept her cool in the sun and dry in the rain.

One day the little old woman was busy baking. She was making round, brown cottage loaves.

She mixed the flour and the water in a big round bowl. She stood at the old wood table, and kneaded the squashy dough.

Knock-knock! Someone at the door ... A man – old and thin. 'Please,' said the man, 'may I have something to eat? I haven't eaten all day. I'm hungry.'

The old woman looked doubtful. Perhaps she felt worried that if she gave some of her bread to the man, she wouldn't have enough for herself.

'Please,' said the man. 'I haven't any money to give you, but whatever you wish for, you shall have.'

So the old woman broke off a little piece of dough and put it in the oven to bake. It puffed up and plumped out and grew and grew, and when she took it out of the oven, it was round and brown – a fine cottage loaf. The warm scent made the old woman think of afternoon tea, with fresh rolls dripping with melting butter. The old woman didn't want to give such a fine, fat loaf away.

So, she broke off an even tinier bit of dough, and put it in the oven to bake. But it was the same again. The dough puffed and plumped and grew and *grew* – into a fine, fat loaf. Much too good to give away!

So this time the old woman took the teeniest, tiniest, wee little *pinch* of dough, and put it in the oven to bake.

But again, it puffed up and plumped out and grew and *grew*.

So, in the end, the little old woman found a dry old crust, at the back of the cupboard, and she gave *that* to the old man. And he took it, and said thank you, and went on his way.

When the old woman was left all alone, she looked at the three fat loaves. And she thought how thin, and how hungry, the old man had looked, and she began to feel bad. 'Oh,' she said to herself, 'if only I could give him a good loaf. But how will I find him, out in the wood? I wish I were a bird, then I could fly up and see the whole forest.' Well, no sooner had she wished than – before you could wink – she began to shrink ... Her arms began to stretch ... and her feet turned into claws ... she was a bird! And – *whoosh!* – the wind whisked her up the chimney.

Up the chimney she went until – *pop!* – out she came, and off she flew. A woodpecker, with feathers of black and white, and the little red hat still on her head.

I don't know if she ever found the hungry old man, to give him some bread, but I do think she learnt something from him. Because, one stormy day, when lightning flashed fire, she met a gentleman woodpecker, soaked to the skin in the rain. And she gave him her best red hat, to keep him dry.

And if you ever see a woodpecker gentleman, you'll see he's still wearing a smart, red hat.

FLIGHT PATTERNS – RECOGNISING A BIRD FROM AFAR

Imagine a bird flying across the sky, and leaving a coloured trail in its wake, which marks its path. The shape of this path through the air is known as a bird's 'flight pattern'. Different birds fly in different ways, and so have different flight patterns.

Knowing a bird's flight pattern can help you to recognise it, even if it's far away.

There are four main types of flight:

- **Flapping**: Wings beat up and down. Some birds, like the Grey Heron, flap slowly. Some flap fast – a Kingfisher flaps so fast its wings are a blue blur.
- **Gliding**: Wings are held out stiff to catch air currents. Gliding uses less energy than flapping, so a bird can stay aloft for longer.
- **Soaring**: Like gliding, wings are held out stiff, and the bird hitches a ride on an updraught of warm air, spiralling upwards into the sky.
- **Hovering**: The bird stays in one spot in the air, by angling its body in the wind and flapping very quickly. The Kestrel is one of the few birds that hovers, as it scans the ground for a movement in the grass or the trail of a vole.

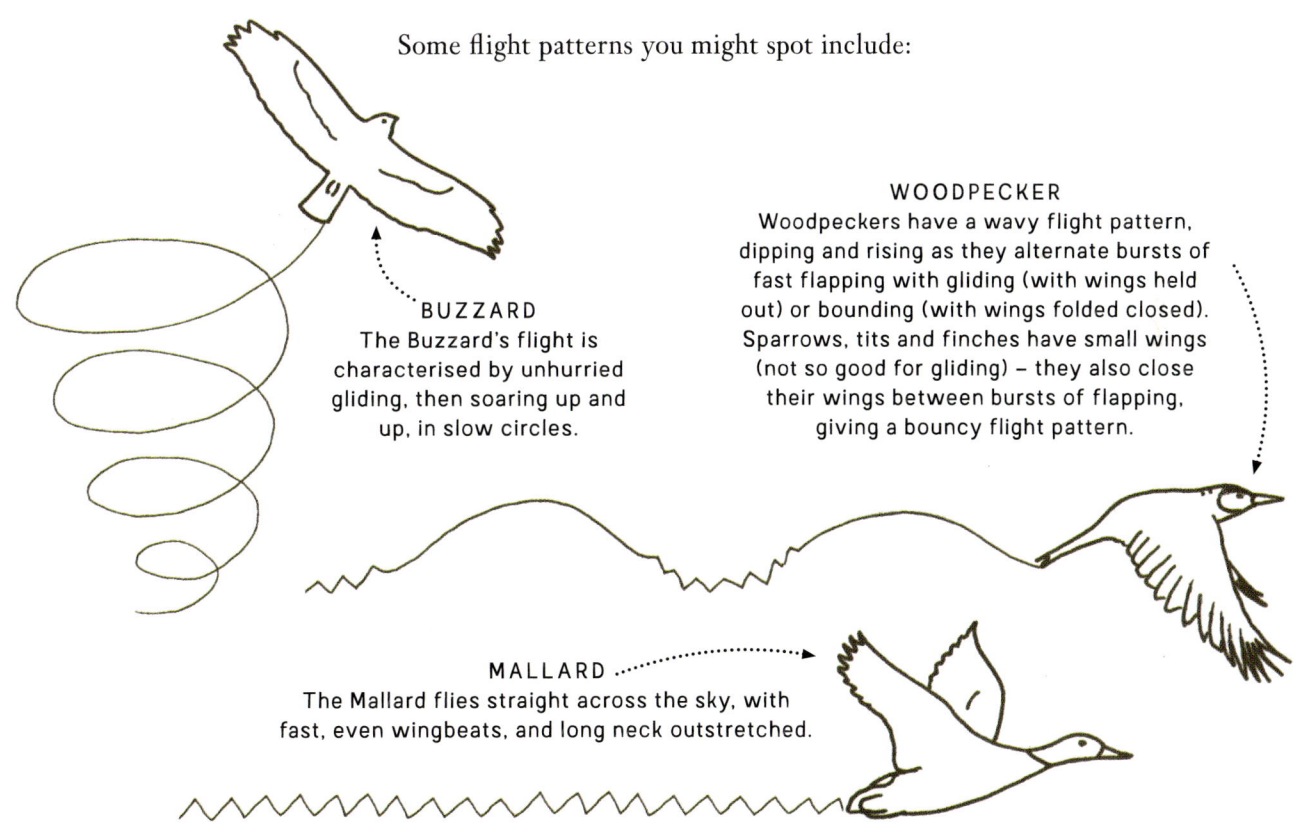

Some flight patterns you might spot include:

BUZZARD
The Buzzard's flight is characterised by unhurried gliding, then soaring up and up, in slow circles.

WOODPECKER
Woodpeckers have a wavy flight pattern, dipping and rising as they alternate bursts of fast flapping with gliding (with wings held out) or bounding (with wings folded closed). Sparrows, tits and finches have small wings (not so good for gliding) – they also close their wings between bursts of flapping, giving a bouncy flight pattern.

MALLARD
The Mallard flies straight across the sky, with fast, even wingbeats, and long neck outstretched.

GREAT SPOTTED WOODPECKER

September balances between the last of summer's warmth
and light and the first of winter's wind and rain. Mornings can
be grey and chilly – spiderwebs sparkle with dew. Afternoons are
golden with the mellow warmth and yellow light of late summer.
On the trees there are yellow leaves now, alongside the green. For
families, it is back-to-school time. In nature too, trees are letting
go of their seed-babies, and many young birds are leaving their
parents' territories to make their own way in the world.

At September's end, we might hear the sound of geese, and look
up to see a skein of birds patterning the sky. In the farming year, the
traditional meal at Michaelmas, on 29th September, was goose.

At this time of grey and gold, we celebrate a golden bird,
and a grey one – the Goldfinch and the Greylag Goose.

SEPTEMBER

GOLDFINCH

Carduelis carduelis

Black crown

Brown back

Yellow-gold
wing-bars

In flight, the broad gold
wing-bars can be seen
in all their glory

Red face

White
cheeks

Black-and-
white tail

Red patch extends behind eye only on male birds.
Juveniles have no black, white or red on heads.

 VOICE
Goldfinches make a delightful
tinkling call, *tick-a-lit, tick-a-lit, tick-a-lit* (or, perhaps, *pippity-pip, pippity-pip, pippity-pip*) as they cross the sky with
a bouncing flight. Listen for the same
twittering chitter-chatter as they feed
together. The Goldfinch's song is an
extended version of the call.

 SIZE
12cm

 WHERE
Gardens, weed-
rich ground.

 NEST
Made of moss,
wool or down,
in tree. Female
incubates eggs.
Both parents
feed nestlings.

 EATS
Seeds, insects.

GOLDFINCH – THE OLD NAMES

From the Goldfinch's colours:
Goldfinch/Goldie (from Anglo-Saxon *Goldfinc*) • **Redcap** (Shropshire, Yorkshire)
King Harry Redcap/King Henry – in Tudor times, a Goldfinch's regal plumage was thought
similar to the royal attire of King Henry VIII • *Lasair-choille* – 'flame of the wood' (Gaelic)
Proud Tailor – the Goldfinch's coat is as grand as the livery of a court tailor (Derbyshire,
Nottinghamshire, Leicestershire, Somerset, Northamptonshire, Warwickshire)

From the Goldfinch's fondness for thistles:
Thisteltuige – meaning 'thistle-tweaker' (Anglo-Saxon) • **Thistle Finch** (Stirling)

From the bird's sweet song: **Sweet William**

THISTLE FINCH, TEASEL TEASER

Seeds are the staple diet of all finches. So, all finches have stout, triangular beaks, which are especially good for tweaking seeds out from seed heads.

The Goldfinch has a larger and more sharply pointed beak than other finches, and can reach into prickly plants like thistles and teasels for seeds. To feast on thistles, a Goldfinch clings to the plant, pulls out the thistledown, bites off the fluff, and swallows the seed. Goldfinches love thistle-seed, and also enjoy seeds from plants in the same family, such as dandelion, burdock and ragwort. Let dandelions live in your lawn, and the golden flowers of spring could bring the golden birds of late summer.

Teasels are also a favourite food. If you see Goldfinches eating teasels, they're probably males. Males have slightly longer beaks than females, so they can reach between the long spines of teasels to get to the seed. This leaves plenty of thistles for the females to enjoy. Empty teasel heads can be used to make a bird feeder especially for Goldfinches. Simply refill an empty teasel with tiny seeds (such as nyjer).

Goldfinches also enjoy the seeds of late-summer flowers such as aster, goldenrod and Michaelmas daisies. So look for Goldfinches in the garden, as well as in open spaces where thistles and teasels thrive.

A CHARM OF GOLDFINCHES

Goldfinches are sociable birds and are rarely seen alone. Pairs often nest close together and work as a team, foraging as a flock.

Goldfinch babies are born in August, when the thistle plants go to seed. After nesting, Goldfinches gather in small flocks and family parties of 6–12 birds. In late summer, watch for the sudden whirring arrival of Goldfinches as they descend on the first seed heads of thistles, with much fluttering and twittering. Listen to the way they buzz and tinkle to each other as they flit between patches of thistles with airy dancing flight.

During the winter, Goldfinches gather in flocks of up to about 40 birds, roaming together between feeding sites and roosting together at night.

The collective noun for a flock of Goldfinches is a 'charm'. The word comes from the Middle English *charme* (and Latin *carmen*), meaning a magic song or spell (the modern meaning of 'charming' comes from the same root). The sound of Goldfinches tinkling together, like tiny golden bells, has charmed us through the ages.

FROM CAGE BIRD TO FREE BIRD

In Victorian times, it was all the rage to keep birds in cages. With their brilliant colours and pretty songs, Goldfinches were popular 'cage birds'. Huge numbers were trapped with nets, and twigs covered with sticky 'bird-lime' (made of mistletoe berries). So many Goldfinches were trapped that by the late nineteenth century they were almost extinct.

In 1889, frustrated that the male-only British Ornithologists' Union were not taking action to protect endangered birds, a philanthropist called Emily Williamson created the Society for the Protection of Birds. The society made saving the Goldfinch one of its first priorities. Now, the Goldfinch is still protected by law and is a common sight. Emily's all-women movement was later awarded a Royal Charter, becoming the Royal Society for the Protection of Birds (RSPB).

GOLDEN BIRD IN A GILDED FRAME

In Renaissance paintings of the Madonna and Child, the Goldfinch is the most widely painted bird after the white dove. Raphael's *Madonna of the Goldfinch*, for example, shows the baby Jesus reaching out a chubby hand to stroke the feathers of a Goldfinch.

Because of its connection with thorny plants, the bird was associated with the crown of thorns Christ wore at the crucifixion. One legend says that the Goldfinch, feeling pity for Christ, plucked out the thorns – a drop of blood splashed the bird's head, giving him the bright red patch he still has today. The golden bird was a shining symbol of resurrection – the miraculous way that death is followed by rebirth, as life springs eternal.

GOLDEN NOTES

The Goldfinch is beautiful to both the ear and the eye. The bird inspired Italian musicians, as well as artists. The composer Vivaldi named his flute concerto *Il Gardellino* after the bird. The light and lively notes of the flute are inspired by the dancing song of the Goldfinch.

GOLDEN RICHES

At the beginning of the seventeenth century, the word 'goldfinch' was used to mean a guinea – a 22-carat gold coin. By the beginning of the eighteenth century, the word had a new slang meaning; a 'goldfinch' meant somebody extremely wealthy (possessing many 'goldfinches'). Hence the Goldfinch's role in the marriage of Cock Robin and Jenny Wren:

'O then,' says Parson Rook,
'Who gives the maid away?'
'I do,' says the Goldfinch,
'And her fortune I will pay.'

MAKE A SUNFLOWER FEEDER FOR THE GOLDFINCHES

Leave seed heads on plants after the flowers are over, to provide food for Goldfinches. And make a sunflower feeder for them to feast upon:

- In late summer and early autumn, when sunflowers start to die off, cut the flower heads from the stalks. Put them somewhere cool and airy for a few weeks to dry (this makes it easier for birds to remove the seeds).
- Take four pieces of string roughly 60cm long, and tie them together in a knot, near the bottom.
- Sit a sunflower on top of the knot and bring each string up and over the flower. Tie the tops of the strings together.
- Hang your sunflower somewhere cats can't reach.
- Your sunflower bird feeder will also be enjoyed by Greenfinches, Blue Tits, Great Tits and Sparrows.

'GOLDFINCH'S GOLDEN FEATHERS' A FOLKTALE FROM ROMANIA

A story celebrating the Goldfinch's distinctive colours

In the beginning, the birds did not have bright colours or pretty patterns. Each had its own call and its own character, but they were all the same plain grey-brown.

So, Mother Nature called every bird in the land together, to receive their special markings. She had pots and pots of shining colours: dark black and bright white, rich red, deep blue, golden yellow. Mother Nature dipped her brush into the pots and one by one she gave the birds their colours. She gave the Robin his red breast and the Blue Tit his bright cap. She painted the speckles on the Thrush and the stars on the Starling.

Soon the pots of colour were almost empty. But there was one little bird who was not there. Thistle-finch.

Where was he? In the thistles, of course, plucking out the sweet milky seeds. Now, a thistle is a very prickly plant. It has prickly leaves and prickly flowers. Even the stem is covered in spikes. It isn't easy to get to the seeds, especially when the wind blows the thistles about. Thistle-finch was busy holding tight to the spiky stem with his stout feet. Tilting his wings to keep his balance in the buffeting breeze. Biting off the fluffy stuff with his sharp-pointed beak.

A fierce gust made the thistles bend and shake. Oh! Ow! Thistle-finch was pinned by prickles, stuck fast in the spikes. So, though he heard Mother Nature calling, he couldn't answer. He twittered and fluttered but the more he struggled, the more he stuck. Sharp spikes caught his coat and pricked his wings. By the time he wriggled his way free, and arrived at Mother Nature, breathless and ruffled, all the other birds had gone.

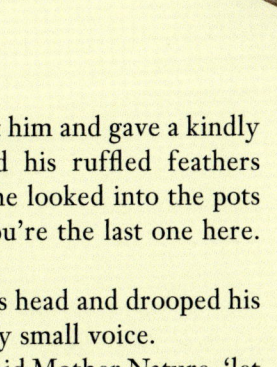

Mother Nature looked at him and gave a kindly smile. Softly, she smoothed his ruffled feathers with the tip of her finger. She looked into the pots of colour. 'Oh,' she said. 'You're the last one here. It's all gone!'

Thistle-finch dropped his head and drooped his wings. 'Oh,' he said, in a very small voice.

'Now, wait a moment,' said Mother Nature, 'let me see …' In just a few pots, right at the bottom, were a few tiny drops of colour. Just a few drops, but in the most bold and beautiful colours – white and black, scarlet and gold.

Mother Nature dipped the tip of her brush. 'Hold still.' She stroked the soft bristles over Thistle-finch's head. He giggled. 'It tickles!'

A dash of black, a flash of white, a splash of red. 'There!' Mother Nature gave Thistle-finch a crown of black, cheeks of white, and a face of scarlet red. 'Now', she said, 'open your wings.' With a sweep of her brush Mother Nature gave Thistle-finch a curve of colour right across his wings. Golden as late-summer sunshine. Thistle-finch was proud as could be of his smart new coat.

And so from that day to this day, the Thistle-finch has had a bright sweep of golden yellow across his wings. Now, he's an expert thistle-tweaker and he never gets caught in the prickles. And he's known by a new name: Goldfinch.

GREYLAG GOOSE

Anser anser

In flight, the leading edges of the Greylag's wings are distinctly pale

Back is grey-brown, and 'barred'

Orange bill

Pale tail

Male Greylags are often slightly larger.

Pale pink feet

VOICE

Hear the clattering clamour of Greylags honking to one another in flight. The cackling *ang-ang-ang* of Greylags is more rattling than the calls of other geese (which are more trumpet-like).

SIZE

75–90cm

WHERE

Open farmland, marshes, lakes.

NEST

On ground, near water, sometimes in colonies. Female incubates eggs whilst male stands guard.

EATS

Waterweeds, crops, grass. Geese are so fond of 'sticky-weed' that the plant is sometimes named 'goosegrass'.

GREYLAG – THE OLD NAMES

Greylag – Some say that 'lag' refers to Greylags 'lagging' behind, when other geese migrate. Some say 'lag' means 'lake'. A 1905 English dialect dictionary states 'lag' is simply an old name for 'goose' – *'lag-lag-lag'* was a phrase used to call or drive geese

Grey Goose – the oldest English name

Wild Goose

From the places the Greylag feeds and breeds:
Marsh Goose • Fen Goose • Stubble Goose (East Lothian)

Geese in flight look like a length of yarn, hence a 'skein' of geese

GREY GOOSE

Grey goose and gander
Waft your wings together,
And carry the good king's daughter
Over the one strand river.

NURSERY RHYME

Geese are notoriously hard to identify. It is a challenge to get close enough for a good look – geese are aware and alert, and so wary of humans that they fly off at the first hint of danger. Hence the saying that describes a fruitless quest: 'a wild goose chase'.

The Greylag is the only native goose that breeds in Britain. There are four other grey geese in Britain, all winter visitors who arrive in late September or October: Taiga Bean Goose, Tundra Bean Goose, Pink-footed Goose and White-fronted Goose. The rarer White-fronted Goose has a white forehead, clearly distinguishing it. To tell the Greylag from the other grey geese, look at the bird's beak. The Pink-footed Goose and both types of Bean Goose all have black markings on their beaks – the Greylag doesn't. A Greylag's beak is large and orange, with a pale tip.

GEESE THAT STAY AND GEESE THAT COME AND GO

The native Greylag was once found as far south as the fens of eastern England. But when the lands it used for breeding were taken for farming, the birds retreated north, to the scattered lochs and heather hills of Scotland.

In the northern highlands (and some Hebridean islands), these native Greylags are resident all year round. Around October, they're joined by Greylags from Iceland, leaving behind Arctic snow and ice. (Some Icelandic birds winter in Ireland too.) For generations, a chorus of honking has been a sound of the season, drawing eyes to the skies for a welcome glimpse of a skein of geese.

In the twentieth century, Greylags were also introduced onto ornamental lakes across England. These domesticated birds are the ancestors of a large population of 'feral' Greylags that once again live wild in many parts of England, and are impossible to tell apart from their native cousins.

THE ANCESTOR OF FARMYARD GEESE

The Greylag is the ancestor of the familiar white farmyard goose. The goose was one of the first creatures to be domesticated by humans – ancient Egyptian paintings show domestic Greylag geese. Geese were easily tamed, as goslings instinctively bond with a parent within hours of hatching, whether the caregiver is a goose or a human. Austrian ornithologist Konrad Lorenz studied this 'imprinting' instinct in Greylags, hand-rearing chicks from the moment they hatched. The geese lived free, and Lorenz went about with a gaggle of goslings trailing after him, and taught them how to fly by running along flapping his arms.

FORMATION FLYING

Skeins of geese, dark 'V' shapes against the bright twilight, make primal patterns in the sky – clean lines and clear arrows. Why do geese fly this way? When a goose beats its wing downwards, the movement causes a swirl of air upwards (just as water splashes upwards, if you beat your hand down into it). This rising splash of air helps lift the bird behind it, making its own flight easier. Geese take turns in the position of leader, so no one bird is exhausted and the extra energy received from flying in formation is shared by the whole flock. Like other geese, Greylags have a bright white 'stern' – a clear flight-flag for others to follow. Flying in formation means no bird is lost from sight (fighter pilots fly in a 'V' formation for the same reason). Geese can even keep in formation in low cloud, when it's hard to see, because they keep honking to stay in touch. If one goose has to land, a few of its family stay with

it until it's able to take flight again – then they all fly off together to find a new flock to join. Geese have much to teach us about leadership, teamwork and communication.

WHIFFLING!

Watch a flock of geese coming in to land and you might spot a goose 'whiffling'. The bird tilts to one side, lowering one wing whilst raising the other. This sky-roll can even twist so far around that a goose is flying for a moment with its head stretched forward as usual, but its neck twisted round by 180 degrees, and its body completely upside down. The bird 'sideslips' – tumbling down through the sky like a fallen leaf, before quickly twisting back round to normal again. The sky-rolling of geese looks like fun (and perhaps it is!) but it's also an expert landing technique. It is thought the twist helps spill air out of the bird's wings, reversing the 'lift' that the shape and position of the wings usually create. Whiffling helps geese brake sharply and descend quickly, avoiding a long slow descent that could make them vulnerable to dangers that speed through the sky, such as birds of prey or the shots of human hunters.

GUARDIAN GOOSE

Watch a flock of geese feeding, and notice how some birds lift their heads and look around – like sentries for the flock, the birds keep watch to allow their fellow geese to feed in safety. If a potential predator comes close to a Greylag's nest, the male bird (gander) will see off the intruder. He then hastens to reassure his mate that all is well – he returns with neck outstretched and loud honks and the pair greets each other by lowering their necks and cackling – a ritual known as 'the triumph ceremony'.

No wonder that, over the years, geese have held the role of guardians for many peoples. In ancient Greece, geese guarded human homes (and like dogs, in some places were companions for children). To the Celts, the goose was a symbol of watchfulness and courage. At the Iron-Age sanctuary of Roquepertuse, in Provence, a huge stone goose guarded the shrine. Goose bones have been found in the graves of Iron-Age warriors in central and eastern Europe and in pagan ship burials in Sweden. In Denmark, too, graves from the Roman Iron Age contained goose bones. Anne Birgitte Gotfredsen, from the University of Copenhagen, says, 'At this time we don't find [geese] in ordinary domestic waste. We only see them as whole animals in connection with rituals or in a grave.' A grave-gift of a goose is thought to have had a double significance; as well as offering spiritual sustenance, the goose was a guardian for the dead.

Caesar wrote that eating goose was taboo amongst Britons, suggesting the bird was held sacred. In Caerwent, in Wales, the remains of a stone statue were found, carved with an inscription to the Celtic god Lenus (identified with the Roman god of war, Mars). All that is left of the statue are two pairs of feet – one pair the feet of a man, the other the webbed feet of a goose. In Brittany, too, a bronze figurine of a warrior-goddess was found, wearing a helmet with a crest in the form of a goose with an outstretched neck.

Roman historian Livy recounted that geese once thwarted an attack on the city with their warning cries (after sentinels and guard dogs both failed to notice any danger). Every year, a giant golden goose was carried through Rome, to honour the birds.

Even today, in Italian vineyards and olive groves, geese not only eat 'pests' (in a pesticide-free approach to growing) but also act as guard-geese.

A LIFELONG BOND

In ancient Greece, geese were associated with the goddess of love, Aphrodite – on Greek tombstones, the image of two geese was a symbol of devoted marriage. If its mate dies, a Greylag usually stays single for the rest of its life.

A pair stays together all year round. Each time they meet, the partners greet each other with the 'triumph ceremony' of neck stretching and cackling, re-enacting their courtship and affirming their bond.

Somewhere near still water, in reeds or a sheltered hollow, the birds build a mound of sticks and vegetation for a nest. The female lines the nest with down, and whenever she leaves the nest to feed, pulls this feather duvet over her eggs to keep them warm.

When their eggs hatch, the female snuggles the chicks close to keep them warm, whilst the male stands guard close by. The yellow goslings follow their parents into the water within hours of hatching. As Lorenz's studies showed, goslings have strong bonds with their carers. Youngsters stay with

their parents all through the winter, even when the geese join with other families in spectacular flocks. Amidst the noisy honking of the flock, geese recognise the voices of their own family.

GRAZING ON GREENS

Three grey geese
In a green field grazing,
Grey were the geese
And green was the grazing.
TONGUE TWISTER

Most geese are grazers – they mainly eat plants. Favourite dining spots are estuaries and fields – especially splashy riverside meadows where water has flooded the fields. Geese like to feed on open ground, which gives them an all-round lookout.

The Greylag's large bill has serrated edges to grip and tear up grass. Geese also pull up reeds and sedges to reach the nutritious stems and roots that grow beneath the surface. A farmer's winter crops are another favourite feast. A clue that geese have been feasting in the fields is crops that are squashed flat – trampled by many large, webbed feet.

A WATERBED

At dusk, geese fly back to safe waters to roost. Open water, such as a lake, gives geese protection from predators like foxes, who are unable (or unwilling) to swim out to catch them. These waterbeds are usually close to feeding areas – wildfowl rarely fly more than five miles between a roost and feeding site.

Wetlands such as marshy bogs and the floodplains of rivers and estuaries are important habitats for geese. In Scotland and northern England, many wildfowl use raised bogs to roost at night and rest in the daytime. A raised bog is created by sphagnum moss. The moss soaks up water like a sponge – as it draws water upwards, it causes the bog to grow upwards, creating raised domes. The process (very slowly) creates peat – it takes up to 1,000 years to grow one metre of peat. Over the last 700 years, many of the peat bogs of Britain were drained for farmland and, more recently, were cut for peat, which was once burnt as fuel. Even today, some compost sold in garden centres contains peat (though it will be banned by 2030).

There are things we can do to help peat bogs thrive. If your garden needs compost, you can make your own. If you buy compost, choose peat-free. If buying plants, choose those grown in peat-free soil. Spot sphagnum moss growing in damp woodland, or on heaths and heather hills, marshes and moors. Each plant is very small, a tiny star of glowing yellow-green or pink-red, and grows close together with many others, creating a spongy carpet of moss (see page 189 for how to access a moss ID guide). If you're out walking somewhere marshy, you can help sphagnum moss find a new home. Break a piece of sphagnum into pieces and sprinkle the pieces in the boggy spot, where they will grow with ease.

A FEATHER BED

Cackle, cackle, Mother Goose,
Have you any feathers loose?
Truly have I, pretty fellow,
Half enough to fill a pillow.
Here are quills, take one or two,
And down to make a bed for you.

NURSERY RHYME

The flight feathers of geese were once used for fletching arrows, as well as for quills (page 131). The soft, fluffy down feathers of geese have been used for bedding for hundreds of years. Our word 'duvet' comes from the French word for 'down'. A feather-filled mattress was known as a 'feather bed'. In olden times it was a comfort that peasants, as well as princesses, could enjoy (especially if they kept a few geese). When geese (or ducks) were being prepared for cooking, the feathers were saved, sometimes over years. Serving-girls were often allowed to keep feathers from birds they'd prepared for the table, and saved them to make pillows or a feather bed for when they were married. As it took so long to collect enough down to fill a feather bed, they were highly valued, and passed down through the generations.

To keep feather beds and pillows fresh and plump, they were shaken and turned every morning when making the bed. A Brothers Grimm folktale tells that when old Mother Holle is making her bed, in her magical fairytale world, here on Earth snowflakes fly like feathers.

Traditionally, feathers were plucked from dead geese that had been farmed for food. In recent times, many commercial makers of pillows and duvets have prioritised profit over animal welfare, plucking feathers from live birds. If buying feather goods, always look for the 'Responsible Down Standard' (RDS) or 'Global Traceable Down Standard' (Global TDS) label.

THE MICHAELMAS GOOSE

There was never such a goose … its
tenderness and flavour, size and
cheapness, were the themes of universal
admiration. Eked out by apple-sauce
and mashed potatoes, it was a sufficient
dinner for the whole family.

FROM *A CHRISTMAS CAROL*, CHARLES DICKENS

Long before the turkey was introduced from America, the traditional Christmas feast was goose.

Christmas is coming,
The goose is getting fat,
Please put a penny in the old man's hat.
If you haven't got a penny,
A ha'penny will do,
If you haven't got a ha'penny,
God bless you.

NURSERY RHYME

In earlier days, the time for goose was Michaelmas, on 29th September. The Christian festival celebrates Michael the Archangel, who overcame the forces of darkness with his blazing sword of light – a fitting picture for the season, to inspire us to keep our own inner light burning bright, as the year turns towards the dark. In the farming year, it was the time that geese, hatched in spring, were ready for market, fat from feeding in the stubble fields.

In Ireland, Michaelmas was known as *na nGéan* – the goose harvest – and in County Cork, the last week of September and the first week of October were known as 'Autumn of the Geese'.

To get geese to the market, a goose girl or boy had to walk the birds to the nearest town. Before their long journey, the geese paddled first through warm sticky tar, then through sand – giving them a pair of strong shoes to protect their feet.

The farmer's wife often kept a flock of geese, and gave them as gifts to friends or those that went hungry. As geese were plentiful in autumn, rural tenants would bring a goose to the landlord when paying their rent – a gift that might make the landlord more forgiving if, instead of cash, an excuse was given at rent-paying time:

And when the tenauntes come to paie
their quarter's rent,
They bring some fowl at Midsummer, a dish
of fish in Lent,
At Christmasse a capon, at Michaelmas a goose,
And somewhat else at New Year's tide,
for feare their lease flies loose.

GEORGE GASCOIGNE, 1575

Eating goose at Michaelmas was believed to bring abundance:

He who eats goose on Michaelmas day
Will never lack money his debts to pay.

There were numerous 'Goose Fairs' in Britain. Tavistock Goosey Fair has been held in October since 1105. Once a weeklong festival for buying and selling geese, it is now popular for stalls and funfair rides, though there are still geese in the livestock market and the town's restaurants offer special menus featuring goose.

WEATHER LORE

Wild geese, wild geese, ganging out to sea,
Good weather it will be.
Wild geese, wild geese, ganging to the hill,
The weather it will spill.

MORAYSHIRE SAYING

In Ireland, if wild geese arrived early, it was a sign that storms and high winds were on the way. Studies show that birds can detect changes in air pressure and that geese may well alter their migrations to avoid an approaching storm, since heavy rain and strong winds make it difficult for them to fly. It is thought that geese detect falling air pressure (a sign of a coming storm) using sensitive receptors in their inner ears.

Perhaps that's why the goose's 'wishbone' was once used to predict the weather (our Christmas tradition of 'pulling the wishbone' is an echo of this old belief). The tradition was first recorded in 1455. On the continent, the goose feast was not at Michaelmas, but at Martinmas (11th November). Dr Hartlieb, physician to Duke Albrecht of Bavaria, recorded the custom: 'When the goose has been eaten … the oldest and most sagacious keeps the breast-bone and allowing it to dry until … morning examines it … Thereby they divine whether the winter will be severe or mild, dry or wet.'

If the November goose-bone be thick,
So will the winter weather be,
If the November goose-bone be thin,
So will the winter weather be.

DIVINE BIRD

In religions across Europe and Asia, geese have a special significance. Engravings at prehistoric settlements of the Old Stone Age show goose-like birds.

In some versions of the Egyptian creation myth, the 'cosmic egg' was laid by a goose – 'the Great Cackler' – a vivid archetype that shines too in the Finnish creation story (page 117), in which the cosmic egg is laid by a duck. Folklore about swans, geese and ducks often overlaps – in early literature and art, words and pictures refer to waterfowl that are not always distinguishable.

In the Hindu epic the *Mahabharata*, for example, the *hamsa* is a sacred bird – the word could be translated as either goose or swan. Either way, the bird is a symbol of the divine and the pure soul. Brahma, the creator god, and his wife Saraswati, a river deity and goddess of wisdom, are both depicted enthroned on the back of the *hamsa*, or flying though the heavens in a carriage drawn by the birds. Brahma himself was born from a 'golden egg, resplendent as the sun', according to the oldest Sanskrit scriptures in Hinduism, the *Brahmanas*.

I wonder if Aesop's fable *The Goose with the Golden Egg* reflects a glimmer of the cosmic egg – it certainly contains wisdom for our times, with its message of the need to care for that which sustains us.

Even today, children enjoy a chocolate egg wrapped in golden paper at Eastertime – a shining symbol of the miracle of life.

In ancient China, too, geese were birds of good fortune and the goose was seen as a messenger between heaven and Earth. The *I Ching*, an ancient book of wisdom used to divine the future, contains 64 'hexagrams' – symbols representing different balances of yin and yang energy. Each hexagram is accompanied by a commentary on its meaning. Hexagram 53 includes two commentaries that emphasise how lucky the goose is, and how sacred:

The wild goose gradually draws near the cliff,
eating and drinking in peace and concord.
Good fortune.

The wild goose gradually draws near the
cloud heights.
Its feathers can be used for the sacred dance.
Good fortune.

'THE GOOSE GIRL' A FOLKTALE FROM CHINA

A story celebrating the way birds inspire humans, and the goose-wisdom of working as one

Once upon a time, there was a goose girl. She was only seven years old, but she knew how to care for the flock. From the morning mist to the golden evening, she tended the geese. She loved looking after them. She loved to watch them gliding over the rippling waters of the duck pond, their white tails reflected in the green-dark water. She loved their stout shapes and soft curves. Grey as autumn cloud, white as mountain snow, their bright beaks the colour of the harvest moon.

She loved the softness of the curling feathers they left upon the shore. She loved to hear them call to one another, keeping in touch with their loved ones. Most of all, she loved the way they flew, making loud music and bold patterns as they flew through the twilight.

'I wish I could fly like you,' said the girl to the geese. 'I wish I could grow wings and fly up into the sky, and see the moon, up close, and look down at all the hills and forests.'

'Well,' she said to herself, 'perhaps I can.'

There was an old well beside the pond and the girl drew up a pail of bright clear water. 'Water helps things grow,' said the girl, and she touched the cold well water to her shoulders, and stood in the sun, and began to beat her arms – practising for when she grew wings.

Well, along came the farmer's daughter. When she saw the goose girl beating her arms up and down, she said, 'What are you doing?'

'I'm growing.'

'Growing?'

'Yes, I'm growing wings so I can fly up into the sky, and see the moon, up close, and look down at all the hills and forests.'

'I don't believe it!' said the farmer's daughter. 'Girls can't fly!'

But when she got home, she was still thinking about flying. What if, perhaps, girls *could* fly?

So the farmer's daughter touched water to her shoulders and she stood in the sun and she beat her arms.

Well, along came the grocer's daughter. 'What are you doing?' she asked.

'I'm growing,' said the farmer's daughter.

'Growing?'

'Yes, I'm growing wings, so I can fly.'

'I don't believe it!' said the grocer's daughter. But when she got home, she was still thinking about flying. Perhaps … girls *could* fly?

She touched water to her shoulders and stood in the sun, and she beat her arms.

Along came the princess of the kingdom herself. When she saw the grocer's daughter, she said, 'What are you doing?'

'I'm growing.'

'Growing?'

'Yes, growing wings, so I can fly.'

'I don't believe it!' said the princess. 'Girls can't fly.'

But even so, the more she thought about flying, the more she wanted to fly.

So the princess touched water to her shoulders and stood in the sun and beat her arms.

And when the girls in the village saw the princess beating her arms, they all stood in the sun, and beat their arms too.

Well, above the high mountain, the Spirit in charge of wings felt the power of their prayers. The reach of their wishes. The air trembling with dreams. And the Spirit flew down to Earth.

The little goose girl felt a rush of air, lifting her up. Her wings beat up and down, lifting her clean up into the sky, and the farmer's daughter, the grocer's daughter, the princess and all the girls of the village followed after her – a long line of girls, flying through the air like geese.

And so the goose girl flew, with a skein of kindred sisters behind her. They could see the moon up close. They could look down at the hills and forests. They could smell the trees and taste the air. They flew through the sky, calling to each other as they went – a whole gaggle of girls, with wings.

WATCH WILDFOWL – VISIT A WETLAND RESERVE

The sight of a skein of geese, stitching the sky with patterns and lines, is a sign that winter is on its way. In towns, ornamental lakes and ponds are good places to watch wildfowl. On the outskirts of cities, reservoirs attract waterbirds. In autumn and winter, geese can be seen on marshlands and wetlands. Estuaries (places where a river meets the sea) offer the chance to see wildfowl in spectacular numbers. When the tide goes out, it leaves a flat expanse of fine mud known as a mudflat. Mudflats support a wealth of life, such as mud-snails and sea-shells – a feast for wading birds. Because mudflats are so open and exposed, feeding birds gather in huge flocks for safety. Many wetlands have a nature reserve where you can watch birds from the cover of a hide. Dawn and dusk are especially good times to visit, to see flocks flying to and from their roost. So, from the end of September, and through winter, wrap up warm, bring a flask and a pair of binoculars, and head outside, to witness a sight of the season – a skein of geese, winging their way through the sunset.

Many organisations can help you find the reserve nearest to where you live. Try:

- The Wildfowl & Wetlands Trust.
- The Wildlife Trusts.
- The RSPB.

GREYLAG GOOSE

September's mellow yellows have deepened into October's
rich orange-browns. Autumn is really here; drifts of fallen leaves
crackle and crunch beneath our boots. Amidst wild wind and rain, we
are sometimes gifted blue-sky days, when the sun lights the colours of
the turning leaves. Beech and bracken are warm brown, rust and russet as
fox-fur and tawny-feather. Hedgehogs collect the dry leaves, ready for
winter hibernation. The nights are drawing in – perhaps, in the
darkness, we might spot a hedgehog or a fox or an owl.

Tawny Owls are rarely seen, but on autumn evenings, amidst the old
trees, we can hear them. It is hooting season, when young Tawny Owls
claim homes of their own. So wrap up warm, head into the dark, and
listen for the fluting, hooting sound of the Tawny Owl.

OCTOBER

TAWNY OWL

Strix aluco

Large round head

Dark triangle on crown

Large black eyes

Clear round facial disc

Rounded body

Tawny-brown plumage, patterned with darker and lighter marks

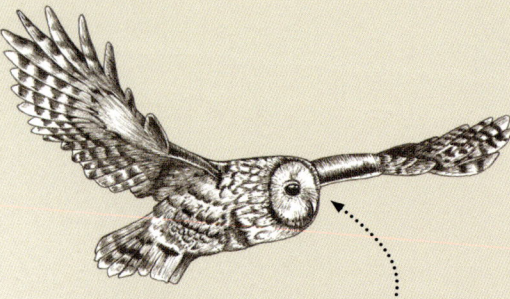

In flight, a Tawny Owl looks rather 'dumpy', with a large head, short tail and broad, rounded wings. Its silent flight is slow and even, with steady wingbeats and quiet glides.

The wings in flight have a distinct bow shape

Juveniles have fluffy pale plumage.

VOICE
A Tawny Owl's wavering song is unmistakable. Often 2–3 short notes are followed by a pause before a final quavering hoot: *hoo-hoo-hoo … hoooooo*. Its contact call is a sharp *kee-wick*.

SIZE
38cm

WHERE
Anywhere with trees. Absent from many British islands, including Ireland.

NEST
Hole in tree.

EATS
Usually small mammals.

TAWNY OWL – THE OLD NAMES

From Anglo-Saxon *Ūle*, from the same root as 'howl': **Owl**

From the Tawny Owl's colour: **Tawny Owl/Brown Owl**

From its habitat: **Wood Owl/Ivy Owl/Beech Owl/Ferny Hoolet**

From the Tawny Owl's hoot:
Hoot Owl (North Yorkshire) • **Hill Hooter** (Cheshire) • **Hoolet/Oolet** (Cumbria, Yorkshire)
Jenny Hoolet/Jinny Yewlet/Gilly Hooter (Shropshire) • **Howlet** (Scotland)
'Ollering Owl (Somerset, Sussex)

Cailleach Oidhche – 'old woman of the night' (Gaelic)

A CALL-AND-RESPONSE LOVE SONG

The Tawny Owl hunts by night – it's rare to see a Tawny. But at this time of year, you might hear one. Tawny Owls hoot in autumn because youngsters, hatched in spring, are now claiming territories of their own. A young male Tawny Owl is especially vocal – his far-carrying hoot proclaims 'I live here!' and, if he's lucky, attracts a mate.

The male shows his mate a number of possible nest-sites, by singing beside each one. The female chooses her favourite – perhaps a hole in an old oak tree.

The familiar hoot of the Tawny Owl, *too-whit, too-whoo*, is actually the sound of *two* Tawny Owls, not one. *Too-whit* (actually a high-pitched *kee-wick*) is a contact call, used to keep in touch in the dark.

Both male and female Owls make the *too-whit* (*kee-wick*) call and the long wavering hoot *too-whoo*. Most often, the female calls *too-whit* and the male responds *too-whoo*. Sometimes, the male sings a long wavering *too-whoo* and the female answers with a *kee-wick*, or with another hoot.

A resident pair sing a duet; the female punctuating the pauses in her partner's hooting song. Their duet tells nearby owls, 'We live here.' Any young Tawny Owl looking for a territory hears the message, and flies on.

NIGHT OWL, SILENT OWL

In the daytime, the Tawny Owl roosts in a tall tree, on a branch close to the trunk or hidden in a hole. By night, the owl hunts. Tawny Owl feathers are velvet-soft, with soft-fringed edges – the softness silences the sound of their wings, allowing the owl to pounce on prey without warning.

Tawny Owls watch for prey from a perch, then swoop down, seizing with sharp claws. At dusk, you may spot a Tawny Owl perched on a branch, TV aerial or chimney pot. On country roads at night, look out for a Tawny Owl perched on a fencepost.

SHARP EARS, SHARP EYES

A Tawny Owl's sharp eyes and sensitive ears help find prey, even on the darkest night.

Owls have forward-facing eyes, which cannot move in their sockets, so to look around they turn their heads instead. Tawny Owls can turn their heads almost full circle to look right behind their own backs.

The stiff feathers that make up a Tawny Owl's 'facial disc' funnel sound towards its ear openings, amplifying the volume. The bird's silent feathers also help it to listen well. A Tawny's hearing is 10 times better than a human's – they can hunt using sound alone to pinpoint a moving target in total darkness, or under a blanket of snow.

Even the rustling of wind or patter of raindrops can muffle the faint scuffles and squeaks owls rely on to catch mice and voles. On wet nights, Tawny Owls don't call or come out, and long periods of rain, especially the crashing noise of heavy downpours, can prevent owls from hunting and even lead to starvation.

Tawny Owls can live for up to 10 or even 20 years, and often mate for life, staying in the same territory and using the same nest-site. Knowing their own patch so well also helps them to find their way in the darkness.

OWLETS

The female Tawny Owl lays her eggs in two batches, with a gap of a few days in between. Each egg takes about a month to hatch, so a family of young owlets includes big brothers or sisters alongside little ones. With its oval shape, woolly plumage, and large dark eyes, the owlet has a rotund look – a tubby ball of fluff.

On summer evenings, listen for their wheezy, squeaky, two-note call, *kee-sip*, as they wait in the crown of a tree, calling continuously to their parents for food. The adults find their young by following their calls. Like a grown-up cutting up a child's meal, they tear up meat and feed the owlets tiny morsels. Tawny Owl parents are fierce protectors of their young – bird ringers wear crash helmets to protect their heads when ringing baby Tawny Owls.

Young owlets leave the nest well before they can fly, exploring the surrounding branches by climbing, jumping, waddling and fluttering. Sometimes, owlets spend time on the ground too – with a bit of effort, they can climb back up to their treetop using their sharp claws. A young Tawny Owl can even scare off a squirrel – it fluffs up its plumage, spreads its wings wide, and *snap-snap-snaps* its bill.

It takes a long time to learn to hunt in the dark. Owlets are cared for by their parents for nearly three months. But by August, they've started hunting for themselves, and by the end of autumn they're independent.

WISE OWL, WISE WOMAN

In ancient Greece, the owl (in particular the Little Owl, *Athene noctua*) was sacred to Athena, goddess of wisdom. The city of Athens was named after Athena, and when people saw an owl in flight, they took it as a sign that the goddess was protecting their city. A wise owl even appeared on Greek coins. To see an owl was a good omen – the saying 'there goes an owl' meant there was a clear sign of victory.

In the Welsh collection of myths and legends *The Mabinogion*, King Arthur's knights seek advice from the Owl of Cwm Cowlyd – one of the oldest of the ancient ancestors. In later centuries, the ancient creatures of the myth featured in the Welsh folktale *The Ancients of the World*. Lonely Eagle wishes for a wife. Not a flighty, feather-brained young thing, but a wise old bird, like himself. His quest leads him to Owl, and the two old birds are happily married.

The wise old owl is a beloved character in many children's classics, such as *Winnie the Pooh*, and often takes the role of teacher, as in Elsa Beskow's *Children of the Forest*. Even today, Brownies often call their teacher Brown Owl.

The owl is a bird of the night, the time of the moon. The cycle of female fertility has also long been associated with the cycles of the moon. In the south of France, if an owl called from the chimney of a woman's house, it was a sign the woman would be blessed with a daughter.

Owls were also thought to be the familiars of witches. Macbeth's three witches add an owlet's wing to their magic brew. We know now that many women of the past who were labelled as witches were herbalists, healers, midwives – wise women.

BIRD OF THE DARK

The owl's love of the night also meant many cultures connected the bird with the idea of death, in traditions where light and dark were associated with life and death. The hoot of an owl spooked some who heard it – especially the shrill shriek of the Barn Owl.

The Romans believed the owl was sacred to Hecate, goddess of the underworld. To see an owl in daytime was considered an omen of death. Indeed, an owl may well mean death for some small creature of the forest – but life for a family of owlets.

MOBBED

Tawny Owls sometimes eat small birds, as well as small mammals. So if small birds discover a Tawny Owl roosting in the daytime, they may 'mob' it with loud alarm calls and even dive-bombs. The placid owl, however, usually ignores the disturbance – it doesn't give a hoot!

Traditional tales offer many reasons for why the owl is mobbed by (or hides from) small birds. In the Welsh folktale *King of the Birds* (page 187), Wren sneaks a ride on Eagle's back. Some versions say the birds cried a pot of tears – planning to dunk the cheeky Wren. Sleepy owl knocked over the pot, and is still scolded for it today. The Brothers Grimm version tells that whoever descended deepest into the earth would be crowned. Wren wiggled into a mousehole – Owl was set to guard the entrance but fell asleep, and Wren flew free.

Welsh myth tells that Blodeuwedd was turned into an owl after trying to kill the hero Lleu, and was harassed by the other birds forever after.

A French folktale tells that Wren flew to the sun to bring back fire. When her feathers were burnt, all the other birds gave her one of their own, except Owl. That's why Owl only comes out at night, when other birds are fast asleep.

'FARMER TICKLE AND THE OWL' A FOLKTALE FROM BRITAIN

A story celebrating the distinctive hoot of the Tawny Owl

Late one night, when old Farmer Tickle was coming home through the dark wood, he lost his way. There was no moon and he wandered round and round, back and forth, tripping on brambles, squelching through bog. He didn't know how to get back and he didn't dare to go on, for fear of sinking in the black bog.

'Hello?' he called. 'Anyone? Help!'

Straight away, a voice answered, 'Whoo? Whoo?'

'It's Farmer Tickle – I'm lost in the wood!'

The voice came again: 'Whoo? Whooo?'

'Farmer Tickle!' he shouted. 'I'm lost!'

Again, the voice: 'Who? Who … who … whoo-o-o?'

''Tis Farmer Tickle, I tell 'ee!' In frustration, the farmer swung his stick, and a great dark shape rushed out from between the trees. And then Farmer Tickle knew just what it was that was tricking him in the dark. 'A pixie!'

The next day, home safe at last, Farmer Tickle told the village folk of his adventure. They all agreed, pixies can be very tricksy.

HOOT LIKE AN OWL AND TALK TO A TAWNY

On autumn and winter evenings, it's magical to go bird-listening. Take a torch with a red light that won't wake the birds or spoil your night-vision. Tawny Owls call from mature trees, defending the area with their hoots, moving from one perch to another as they hunt. Places with large, old trees, like churchyards or woodlands, offer the best chance to hear owls call.

Choose a calm, windless night (Tawny Owls call more often on dry, still nights, when their hoots carry 69 times further). Be still and listen. If you hear an owl call, you can hoot in response. If you're patient, the owl might hoot back. A Cambridge-shire study found that in over 90 per cent of trials, human 'owl calls' received a response from an owl within half an hour.

Wordsworth's poem 'There Was a Boy' cele-brates the joy of talking to a Tawny:

And there, with fingers interwoven, both hands
Pressed closely palm to palm and to his mouth
Uplifted, he, as through an instrument,
Blew mimic hootings to the silent owls,
That they might answer him.—And they
would shout
Across the watery vale, and shout again,
Responsive to his call,—with quivering peals,
And long halloos, and screams, and echoes loud
Redoubled and redoubled; concourse wild
Of jocund din!

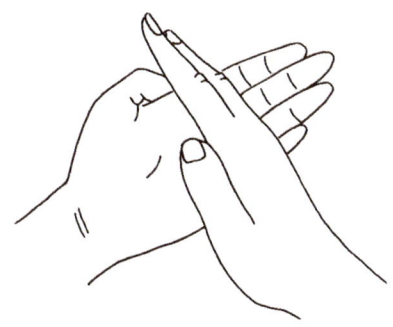 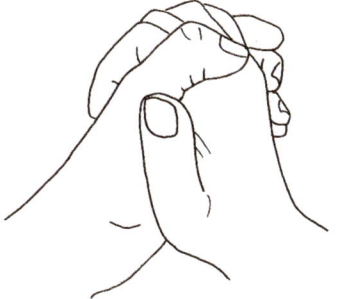 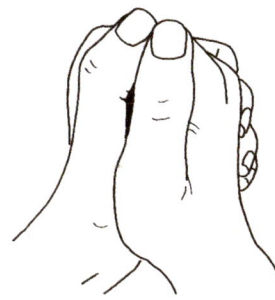

1 **Cup your hands:**
Place one hand over the other so the left edge of your right hand is lined up with the base of the fingers of your left hand. Now pivot your right hand, where the hands touch, so that it sits face down on top of your left hand. Curl the fingertips of each hand around the other hand. Make a space between your hands by widening the clasp of your fingers. Don't leave any gaps where air could escape.

2 **Line up thumbs:**
Keeping your hands together, move your left thumb so that both thumbs are aligned, upright and pressed together. Bend the top part of your thumbs over a little, away from you.

3 **Breathe and blow:**
Purse your lips a little. Bring your hands up in front of your mouth. Press your lips against the top joint in your thumbs then blow through and down into your hands.

Keep practising until you can hoot like a Tawny Owl.

The bright blaze of October has dimmed, like a lamp turned low at the end of the evening. The last leaves glow, like the embers of the fire. Day after day the winds and the rains return the fallen leaves, the energy of the old year, back into the earth; compost for next year's growth.

Trees let go of their leaves and their seeds – death and birth nourishing life. Old leaves and new seeds bed down together in the dark earth. The soil is made of generations of trees, plants, animals and birds. It is a time to remember those who have lived before. We look back and honour our ancestors.

And, perhaps, we begin to look forward too, to midwinter gatherings with those we love. The birds are also preparing for the winter. Summer visitors have already flown. Many others are gathering together into winter flocks. Starlings gather together in the largest flocks in Britain.

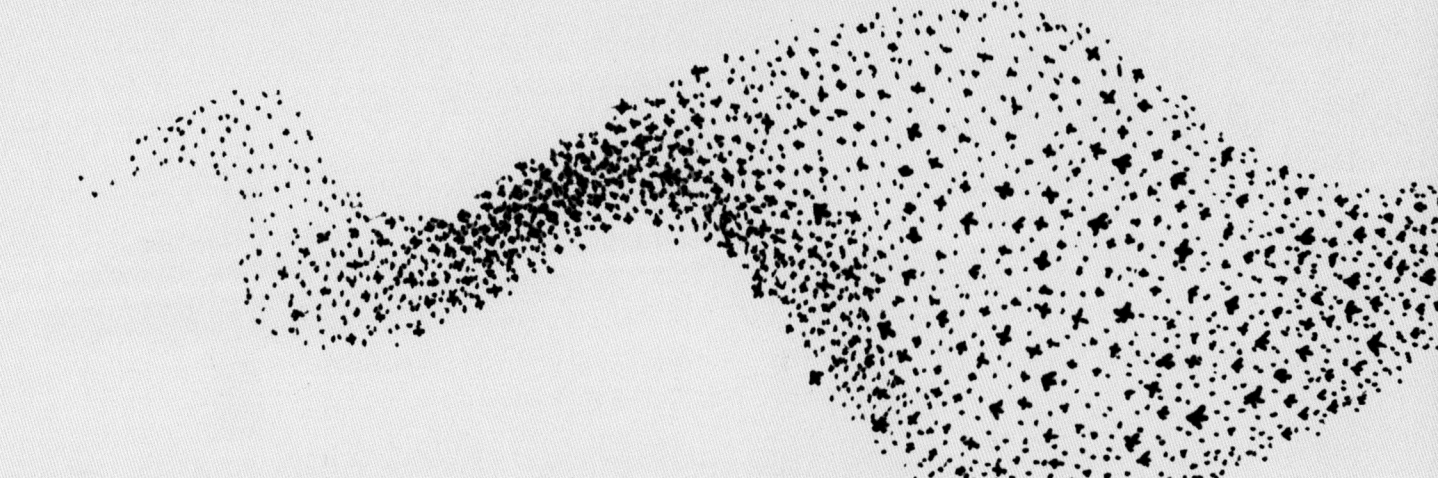

NOVEMBER

STARLING

Sturnus vulgaris

Pointed bill ·········

White speckles

Glossy black ················
feathers have a
purple-and-green
sheen

In winter and spring,
the base of a male
Starling's bill is blue,
whilst the female's
is pink. Juveniles are
grey-brown. The young
bird's head feathers
are the last to change
colour – by September
its grey-brown head is
all that distinguishes
it from the adult.

Seen from below, a Starling's
triangular wings, pointed head and
short tail give a distinctive silhouette,
like a four-pointed star

MALE – SUMMER
PLUMAGE

VOICE
Both male and female Starlings
sing throughout the year, a noisy
jumble of whistles and clicks,
rattles and trills, squeaks, squeals
and scratches. Starlings are expert
mimics – if you hear a song that
sounds like a phone, it's a Starling.

SIZE
21.5cm

WHERE
Near fields.

NEST
Hole in tree/
building. Male
builds nest,
female lines with
feathers/moss.
Both parents tend
eggs and chicks.

EATS
Insects,
earthworms,
spiders, slugs,
fruit, seeds,
roots.

STARLING SOUNDS

The Starling is an expert mimic, the best of all British birds. It can copy the calls of other birds, including Tawny Owls, Buzzards and Green Woodpeckers, and imitate animal noises, including hens, ducks, sheep, cats and frogs. Mechanical sounds, like telephone ringtones and car alarms, are also part of its repertoire. A Starling can even mimic human speech.

In a story from *The Mabinogion*, the heroine Branwen escapes her abusive husband with the help of a Starling. She teaches the Starling to speak and the bird carries her message to her brother Brân, who journeys across the sea to rescue her.

Mozart, too, famously had a pet Starling. He first heard the bird in a pet shop in Vienna – it was whistling one of his own compositions. How the bird had heard the tune was a mystery, but Mozart was enchanted, and the bird became his beloved pet. Some have speculated that Mozart's piece *A Musical Joke*, with its free form and 'jokey' moments (discordant horns and jumping key changes) was inspired by the glorious jumble that is a Starling's song.

Weaving together the sounds of their surroundings, Starling song is a sound-map of the land. In chorus, Starlings' high-pitched chatter rises above the volume of city traffic. The sound of a party of Starlings is a noisy, non-stop cacophony of scrape and squeak, click and rattle, buzz and wheeze and whistle.

Starlings often accompany their singing with energetic wing-waving, a sight that can distinguish a Starling at a distance.

THE STAR-SPANGLED STARLING

The Starling's plumage is dark as a winter's night. Each dark feather is tipped with a fleck of white – the bird's dark winter coat is spangled with bright stars. As the year turns, the white feather tips gradually wear away, revealing fine new feathers beneath. By spring, the breeding season, the Starling's glossy black plumage gleams with blues and greens, a courting coat of shot silk.

FLOCK TOGETHER

As winter approaches, many birds flock together. Sociable Starlings come together, sweeping the sky in swooping swarms known as 'murmurations'. These are the largest flocks in Britain – including thousands, sometimes millions, of birds. The sound of so many Starlings is as loud as an express train. The number of Starlings almost doubles in winter,

when resident birds are joined by visitors from eastern Europe and Russia. For us, the sight of a murmuration of Starlings is a winter joy, while in Russia, the bird is a herald of spring, as well as a farmer's friend. In communist times, Starlings were provided with 22.5 million nest boxes, all looked after by schoolchildren as part of their state education.

Chattering flocks of Starlings can often be seen perched on telephone wires or power lines. Why do birds flock together? A solitary bird needs to keep a constant lookout for danger, but in a flock, some birds stand guard whilst others eat. Birds in a flock spend less time looking around and more time feeding.

When trees are bare and there are fewer places to shelter from predators, there's safety in numbers too. Flying close together makes it harder for birds of prey to target an isolated bird, especially when the flock flies as one, moving and turning in tight formation.

FLOCK TOGETHER, FEED TOGETHER

In autumn, farmers' fields are a good place to spot Starlings feeding. They love to feast on leatherjackets, the grubs of the daddy longlegs (or crane fly). In winter, too, look for Starlings in fields of sheep or cows. They land right on top of a sheep's back to pick off ticks – the sheep receive some haircare and the birds receive a bug-buffet. Animals' feet also disturb the ground, making insects easier to find. Watch how a Starling finds food in the earth. The bird probes into the turf with its beak, then prises the hole wider by opening its bill in the ground. This 'prying' allows the Starling to see if there's a grub in the hole. Starlings are the only British birds that peck for worms and grubs with open beaks.

Despite their ingenious feeding techniques, Starling numbers fell by more than half between 1995 and 2018 – the bird is currently 'red listed'. Recent farming practices, where pesticides eradicate Starlings' food, may be one cause – we can foster life-sustaining farming by supporting regenerative agriculture and buying organic food.

FLOCK TOGETHER, ROOST TOGETHER

At the end of the day, as commuters flock out of the city, Starlings flock in. Towns are usually warmer than the surrounding countryside, so Starlings often come to city centres to roost, flying up to 30 miles from their rural feeding grounds. In the 1940s, Starling flocks roosting in the centre of London were so huge that when they landed on the hands of Big Ben, they stopped the clock.

During the day, Starlings gather in feeding parties. As the sun sinks, different bands of birds fly in from all directions, meeting high in the sky and joining together. At the communal roost, a reed bed perhaps, or an old pier, the chatter of Starlings is noisy and non-stop. There are animated discussions with neighbours as they jostle for space: 'budge up … 'scuse me … I'm here … night night.' With so many conversations going on at once, the volume gets louder and louder. Sometimes, just before they settle down for the night, the whole flock bursts back into the sky and we are treated to a spectacular aerial display.

FLOCK TOGETHER, FLY TOGETHER

The dusk dance of a constellation of Starlings is a sight to behold. As one, the vast company dips and rises, twists and turns, swoops and swerves. Like ink swirling, like fish shoaling, like smoke billowing. Like Northern Lights dancing across the sky, the Starlings flow and fold, stretch out and close in, patterning the sky with pulsing life.

The Starlings' sky-dance is perfectly synchronised. How is such precision formation flying achieved? It is thought that when one bird changes direction, those surrounding it do too, then their neighbours follow, so the change of direction ripples through the whole flock. Imagine being able to respond with such fine-tuned awareness and lightning speed!

Why do Starlings perform these aerial acrobatics? Perhaps because they can! Scientists suggest the dance acts as a visual invitation, attracting Starlings from far and wide to join the flock.

In ancient Rome, the shapes formed by Starling murmurations were seen as signs that told if plans were, or were not, in line with the will of the gods, and it was the special role of the *augur*, 'one who looks at birds', to interpret their meanings.

Whatever the meaning of a murmuration, one thing is clear: lit by the winter sunset, a sky full of Starlings is a true wonder.

WATCH A STARLING MURMURATION

There are some well-known Starling roosts in Britain, such as Gretna Green in Dumfries and Galloway and Brighton Pier in Sussex. Several RSPB reserves attract roosting Starlings – investigate the reserves in your local area (page 189). Even if you don't know a local roost, you can watch the skies as the sun sinks on winter afternoons. Starlings often use the same sky-paths each day, just as we might walk the same way home each afternoon.

If you plan to visit a Starling roost, it's best to get there at least an hour before sunset (once the light starts to fade, things quickly quieten down). Wrap up warm. Bring a flask, and snacks for children. Find somewhere to watch that won't disturb the birds. Enjoy the show!

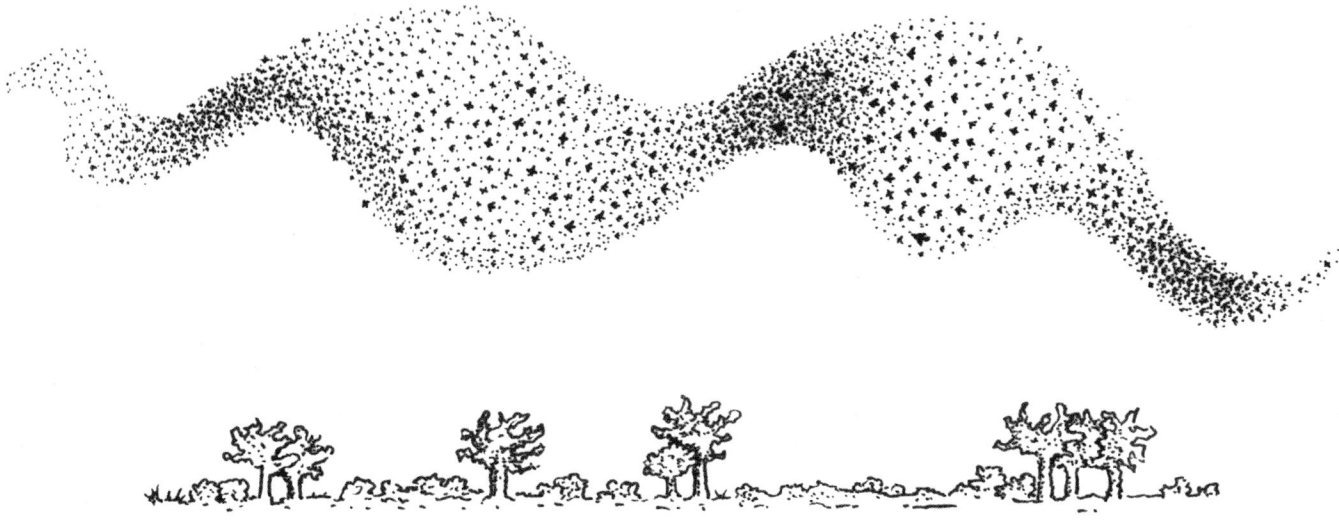

'WHAT BIRD WOULD YOU BE?'
A FOLKTALE FROM ENGLAND
A story that values the way Starlings flock together

Once there was a king, at the twilight of his life. He lay on his deathbed, pondering how best to help his sons after he had gone. He called the three boys to his bedside. 'Tell me,' he said, 'if you could be a bird, what would you be?'

'A hawk!' said the first son. 'The noble bird of the valiant knight.'

The king nodded. He looked to his second son.

'An eagle,' said the second son. 'The most powerful hunter, feared by all the birds.'

'Thank you,' said the king. 'And you?' He turned to his youngest son.

The boy took his father's hand. 'I'd be a Starling,' he said. 'When I see the Starlings at sunset, my spirit soars. The way the whole flock flies as one! Each one in connection, each one playing its part in the whole.'

The king smiled. 'Thank you. I see now how best to help each of you.'

When the king was laid to rest, he left his two eldest sons land to live upon. To his youngest son he left no land at all. He saw that the wisdom of the youngest son was all he needed to live well.

And so it was. The first son became a warrior. He lived by the sword and, before long, he died by the sword. The second son seized power over many, but he was feared by all those in his power. The youngest son lived in connection with his kinsfolk. He devoted himself to serving the whole – all life. He was loved by one and all, and, when the time came, the people made him their king. All thanks to the wisdom of the Starlings.

A CLOSER VIEW – HOW TO USE BINOCULARS

Binoculars come in different sizes, such as 8×30 and 10×40. The first number refers to the magnification power of the binoculars – 8×, for example, makes an object look eight times bigger. The second number refers to the size of the front lenses (and therefore how much light they let in). The higher the second number, the more light is let in (which is better for birdwatching in low light). But the larger the second number, the bigger and heavier the binoculars will be. Large, weighty binoculars are harder to carry around.

Generally, a good magnification is 8×. Higher magnifications can amplify natural movements of the hands, giving a shaky image. For general use, a pair of 8×40 binoculars are ideal. Binoculars are easily available second-hand, and pre-loved buys are better for the environment too.

Binoculars need to be focused before use.

- Find a distant, non-moving object that has sky as a background, such as an aerial or dead branch.
- Close your right eye. Use the central wheel to turn the left lens of the binoculars until the object comes into focus.
- Close your left eye and look through the right lens. Use the adjustable eyepiece to make the image look clearer.
- Leave the adjustable eyepiece in place. When you see a bird, use the central wheel to focus.

Getting the hang of binoculars takes a bit of practice. Wear binoculars around your neck (rather than carrying them in the case) so they're to hand when you spot a bird. Often, it's easier to spot a bird first, then raise your binoculars to see it more closely, rather than looking for birds through binoculars. Use your peripheral vision to help you – be aware of any subtle movements that you see in the corner of your eye.

Birds are always alert to any danger, so move mindfully (avoiding sudden movements) so you don't scare them away. You can dress 'quietly' too – in natural colours that blend with natural surroundings and in fabrics that don't rustle when you move.

On a sunny day, keeping the sun behind you helps, as birds are harder to spot in a bright sky (remember too, never to look directly at the sun). Standing behind (or in front of) a tree and peering round it can keep you camouflaged. If there's no cover nearby, simply stand still and stay silent.

Lenses

Central wheel

Adjustable eyepiece

It is midwinter. Days are short. Nights are dark and cold. The mood is deepest blue – there is a stillness, a silence, akin to reverence. In the darkness, light shines brighter – winter stars, glowing fires, golden moments.

In the heart of winter perhaps no other bird is more celebrated than the Robin, whose cheery red breast brightens so many Christmas cards and whose song brightens the winter quiet. The Wren is also a bird of the season – the winter king traditionally honoured in Britain, and beyond, during the twelve days of Christmas.

Both the Robin and the Wren live in our gardens through the seasons. Whatever the weather, the Robin on your spade and the Wren in your hedge will be with you all winter long.

DECEMBER

ROBIN

Erithacus rubecula

Orange-red breast and face

Juveniles have speckled breasts

Rounded shape

VOICE
If you hear a bird singing between autumn and Christmas, it's most often a Robin – the colder months are the perfect time to become familiar with the Robin's twiddly, scribbly song: *twiddle-oo, twiddle-eedee, twiddle-ee.* The Robin's alarm call is a loud, repetitive *tic ... tic, tic.*

SIZE
14cm

WHERE
Gardens, woods, hedges.

NEST
Female makes mossy nest in undergrowth or ivy. Sometimes nests in unusual places (farm machinery, sheds, kettles, boots, hanging baskets, coat pockets). Mate brings female food while she builds. If disturbed, Robins will abandon a nest in-progress.

EATS
Insects, worms, berries.

DECEMBER

ROBIN – THE OLD NAMES

Broindergh – 'red belly' (Gaelic) • *Rudduc* – 'ruddy' (Anglo-Saxon)
Ruddock (Northern England)/**Reddock** (Dorset)

Redbreast (from Middle Ages to 1952, when 'Robin' became official)

Bob Robin (Stirlingshire) • **Bobby Robin** (Suffolk, New Forest)

Spideog Mhuire – 'the Virgin Mary's Robin' (Irish)

ROBIN REDBREAST

A Robin's bright breast attracts females and deters other males. The feathers are orange-red, but no British bird has the word 'orange' in its name, because the colour did not enter the English language until the first orange fruit was brought to England in medieval times. By that time, the 'Redbreast' was already named. Later, it became popular to give human names to familiar birds (in the same way we give friends and family affectionate nicknames) and the bird became known as Robin Redbreast.

ROBIN'S PERCH

A Robin perched on a garden spade is a familiar sight. Robins catch their food using a 'perch and pounce' approach – they perch quietly, not too far from the ground. A Robin's large eyes help it spot insects moving over the earth, even in the shade of woods and thickets, and at first light and last light, when it's too dark for other birds to feed. The Robin scans its patch, waiting, and when it spots a bug, it hops down and snaps it up.

A WINTER SONG

Most birds sing only in the springtime. The sweet song of the Robin, however, can be heard almost all year round. Both male and female Robins sing in the autumn. The Robin's autumn song is quiet and mellow – it is a gentle reminder to other birds of the Robin's feeding territory.

But sometime in December, the male Robin changes his tune – literally. The subdued autumn song is replaced by a more energetic, passionate song – a love song intended to attract a mate. When a female hears the serenade, she comes into the male's territory to investigate. She may visit the territories of several males before choosing a mate. A male Robin defends his territory with bold vigour, and sometimes even rebuffs a visiting female a few times before realising her appeal. The female may join her territory with his, or return to her own patch until breeding begins in spring. During this time, neither bird will change partners – they're committed to each other until they've raised their family together. If you see two Robins foraging in a garden, close together in comfortable calm, they're a pair – you've been lucky enough to witness a Robin's winter love affair.

WINTER WARMTH

Through the cold months, enjoy the Robin's cheering song, and notice how the bird fluffs up its feathers in the wind, to keep warm.

You might even see a Robin perched on one leg – to keep their bare feet warm on chilly days, many birds tuck one leg up into their feathers to stay snug. Or, as the rhyme says:

The North Wind doth blow,
And we shall have snow,
And what will poor robin do then, poor thing?
He'll sit in a barn,
And keep himself warm,
And hide his head under his wing, poor thing.
SONGS FOR THE NURSERY, 1805

FIERY FEATHERS

Many folktales explain how the Robin got its red breast. A French folktale tells how, when the Wren brought fire from heaven, her feathers caught alight, and the Robin came to her rescue, taking the flame. He passed it to the Lark, who brought it safely to Earth, but the Robin's breast has been red as fire ever since. In the folklore of Guernsey, too, it was the Robin who first brought fire to the island from the mainland, burning his feathers flame-red as he carried it over the water. The first record of the tale came from an old woman who said, 'My mother had a great veneration for the Robin, for what should we have done without fire?'

A Greek legend tells us that as a child Jesus fed the Robin outside His mother's door, nurturing a lifelong bond with the bird. When Jesus ascended to heaven, the legend says, the Robin sang with the angels.

A story from Brittany tells that the Robin perched upon the cross, brushing away Christ's tears with her wings, and plucking out the spikes from His crown of thorns. A drop of blood stained the Robin's feathers red. Jesus blessed the bird, saying, 'Wherever you go, happiness and joy shall follow.'

BLESSED BIRD

The Robin and
the Jenny Wren,
Are God Almighty's cock
and hen.

Gaelic lore tells that the Robin is so sacred that even the bark of a briar rose that once held the bird's nest can be made into a healing brew. The Robin was beloved far and wide – to take an egg from such a blessed bird was considered an unthinkable act of sacrilege. In Dartmoor the penalty attached to this crime was the smashing of all the crockery in the house.

To disturb a Robin's nest, or kill a Robin, is sure to bring bad luck:

The robin and the redbreast
The robin and the wren,
If ye take out of the nest
Ye'll never thrive again.
TRADITIONAL RHYME, ESSEX

FEATHERED FRIEND

The Robin's everyday presence and bold curiosity have made it a familiar feathered friend in our own gardens – Robins are Britain's best-loved birds. They are with us all year round, and if we're mindful in our movements (especially when digging, turning up worms) are unafraid of coming close. Many people enjoy a special relationship with their local Robin. In the hope of a worm, Robins once followed in the footsteps of wild boar, and also follow moles and deer (whose large feet and probing noses unearth all manner of morsels).

The down-to-earth appeal of the Robin is expressed in the old nursery rhyme:

Little Robin Redbreast
Sat upon a rail,
Niddle, noddle went
his head
Wag went his tail.

DEATH AND BURIAL OF
COCK ROBIN, 1797

MESSAGE BIRD

Because the first postmen wore bright red waist-coats, they were known as 'robins', and Christmas cards of the time often showed a Robin holding an envelope in its beak – delivering a message of good-will. The bird's bright breast is still a cheering sight on Christmas cards, as well as in the wild.

A sixteenth-century 'Cornucopia' states that a Robin, 'if he find a man or woman dead, will cover his face with moss'. The idea that (as Shakespeare says) the 'Ruddock with charitable bill' covers the bodies of unburied humans with leaves was later made famous in the ballad *Babes in the Wood*. Even today, many people can share a personal story of a Robin that appeared at the sad time of a death, bringing a sense of comfort and connection – delivering a message from the spirit world of lasting love.

A GIFT FOR THE BIRDS

Just as we receive the winter gift of song from the Robin, we can offer the Robin, and all the birds, a gift in return. Winter is the perfect time to feed the birds. Insects and berries are scarce and birds need high-fat foods to stay plump enough to survive cold nights. Give the birds a Christmas present:

You will need:

- Lard/suet
- A mixture of seeds, nuts, dried fruits, raw oatmeal, cheese
- Cookie cutters

1 Melt the lard and pour into a heatproof bowl.
2 Add the remaining ingredients – use about one-third fat to two-thirds of mixture. Stir well.
3 Arrange the cookie cutters on a layer of waxed paper, on a freezer-proof tray.

4 Scoop the mixture into the cookie cutters and pat flat. Use a straw or pencil to make a hanging hole in each shape.
5 Freeze for a few hours to reset the lard.
6 Once set, remove the cookies from the cutters and thread twine through the holes. Hang in the branches for the winter birds.
7 The bird cookie mixture can also be pasted into a coconut half or half an orange with the flesh removed.

Blackbirds and thrushes feed on the ground. In winter, when earth is hard or snow-covered, windfall fruit is vital food. Help ground-feeding birds by putting apples (even old ones) out on the lawn. Space them out, so different birds can feed at the same time. If there's snow, clear a small patch of ground before putting down food.

Put food out regularly, as birds come to depend on it. A guaranteed food and water supply can make the difference between life and death.

'ROBIN'S WINTER GIFT' A FOLKTALE FROM SCOTLAND

A story celebrating the gift of the Robin's winter song

It was Christmas morning. And what a glorious morning it was! The dips and hollows were white with frost and the grass glistened with dew. The beech and the bracken were warm brown in the morning sun and the earth was breathing out mist.

The Robin was so full of joy that he couldn't help but sing. He perched on the branch of a briar and sang his cheerful song. 'Twiddle-oo, twiddle-eedee, twiddle-ee.'

Beside the tangle of bramble ran a peat-brown burn. And by the side of the burn, padding on velvet paws, came a pearl-grey pussycat.

The Robin gave a call of alarm, 'Tic … tic-tic'.

The cat stretched luxuriously. She arched her back and tilted her neck, so the Robin could see her pretty fur. 'Little Robin, Little Robin, where are you going so early this morning?'

'I'm away to the castle, to sing for the king.'

'Mmm … a worthy journey. But, before you go, hop down here. Upon my neck is a ring of bonny white fur. Come closer and see for yourself.'

But the Robin replied, 'No, no, Mistress Cat. For I saw you worry a mouse, and I have no wish to be worried by you!' And away he flew … over the gorse and over the heather …

… until he came to rest on a bank of turf at the edge of a field. He hopped here and there, pecking the ground beside the hedge.

At the other end of the bank, perching on a fencepost, was a hawk. The hawk shuffled along the fence, closer to the Robin. The Robin hopped further away. The hawk looked over his hooked beak. 'Little Robin, Little Robin, where are you going so early this morning?'

'I'm away to the castle, to sing for the king.'

'Ah, a fine idea, but, before you go, hop up here. Upon my wing is a curious feather. Come closer and see for yourself.'

But the Robin replied, 'No, no, Mr Hawk. For I saw you pluck the feathers from a Linnet, and I have no wish to have my feathers plucked by you!'

And away he flew … into a thorny thicket, where the hawk could not reach. From the safety of the hedge, the Robin gave a high, thin cry, 'Tseee' – a warning to all his feathered friends.

When the hawk was gone, the Robin flew on … swooping over thistles, sweeping past sheep … until he came to rest on the edge of a flat rock. From under the rock rose a musky stench, and from out of a hole came a twitching nose … quivering whiskers … two pricked ears …

Fox swung her splendid tail. 'Little Robin, Little Robin, where are you going so early this morning?'

'I'm away to the castle, to sing for the king.'

'Ah, indeed. But before you go, hop down here; at the tip of my tail is a tuft of white. Come closer and see for yourself.'

But the Robin replied, 'No, no, Mistress Fox. For I saw you catch a wee lamb, and I have no wish to be caught by you!' And away he flew.

The robin flew from oak to ash to thorn, until he came to the grey stone castle of the king. He lit upon the windowsill. But there, in the ivy underneath the window ledge, someone was already singing! Someone round and brown with a turned-up tail – little Jenny Wren. The Robin greeted the Wren with a nod and a bow. The Wren bobbed a curtsey.

Together the Robin and the Wren sang for the king and the queen – a thrilling trill and a twiddly tune. The king and the queen came to the window to listen, and the singing of Robin Redbreast and Jenny Wren made their hearts rejoice.

MAKE A ROBIN PICTURE

Children can follow these simple steps to make a drawing of a robin.
It could be made into a Christmas card for someone special.

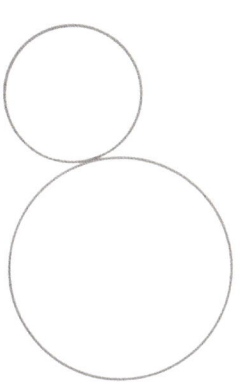

1 Using a pencil, lightly draw two circles, one big and one small (like a wonky snowman).

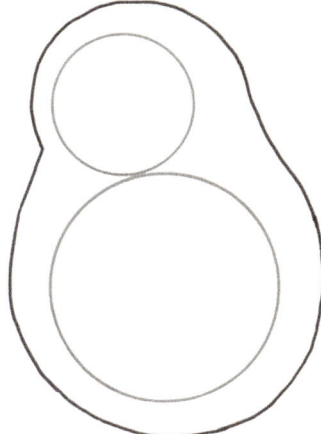

2 Draw around the circles to make the shape of the Robin's body.

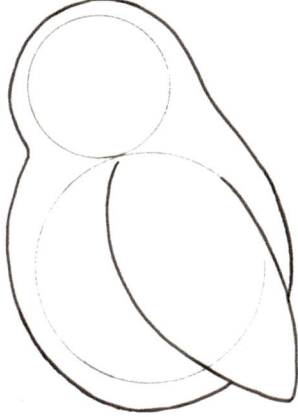

3 Starting at the top of the big circle, draw a wing that ends below the body.

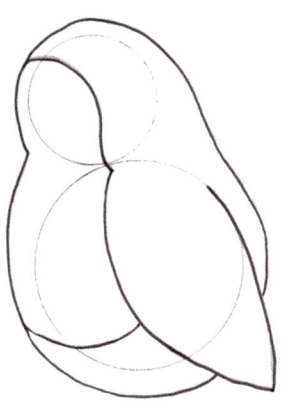

4 Draw in the shape of the Robin's red breast. If you like, rub out the pencil lines of your original two circles.

5 Add an eye, beak, feet and tail.

6 Colour your drawing.

WREN

Troglodytes troglodytes

Warm brown

Round shape

Short tail sticks up

Buff below

Darker bars

VOICE

The Wren's repeated *tit-it-it* call can intensify into a rapid chittering scold. The call has a metallic tone, like clockwork winding down in irregular bursts. The Wren's song is exuberant: a loud volley of clear, shrill notes mixed with (and often ending on) intense rattling trills.

SIZE

9.5cm

WHERE

Low down.

NEST

In sheltered nook. Both parents feed young.

EATS

Insects, spiders.

WREN – THE OLD NAMES

From Anglo-Saxon *Wrænno* ('lascivious' – from Wren's numerous children): **Wren**
Wranny (Cornwall) • **Wrannock** (Orkney Isles)

From the Wren's short tail: **Cutty/Cut** – from Welch *cwt*, 'short tail' (Dorset, Devon,
Hampshire, Pembrokeshire) • **Cutteley Wren** (Somerset) • **Scutty** (Sussex)
Bobby Wren – i.e. bob-tail (Norfolk) • **Stumpy Toddy** (Cheshire) • **Stumpit** (Lancashire)

From the Wren's tiny size: **Tiddy Wren/Tidley Wren** (Essex)
Titty Wren (Devon–Suffolk) • **Titmeg** (Devon)

From the Wren's rattling voice: **Crackadee/Cracket/Crackeys/Crackil** (Devon)

Pet names: **Jenny Wren** • **Sally Wren** (Ireland)

In many European lands, the Wren has a royal or divine title: **Our Lady's Hen** (Scotland)
Oiseau de Dieu – 'Bird of God' (Normandy) • *Regulus* – Little King (Latin) • *Roitelet/
Roi des Oiseaux/Roi de Froidure* – Kinglet/King of Birds/King of Coldness (French)
Zaunkönig/Schneekönig – Hedge King/Snow King (German)

LITTLE BROWN MOUSE, LITTLE BROWN WREN

Of all the birds that rove and sing
Near dwellings made for men,
None is so nimble, neat and trim
As Jenny Wren.
With pin-point bill, and tail a-cock,
So wildly shrill she cries,
The echoes on their roof-tree knock
And fill the skies.
Never was sweeter seraph hid
Within so small a house–
A tiny, inch-long, eager, ardent,
Feathered mouse.

'JENNY WREN', WALTER DE LA MARE

In the 1940s and 1950s, the Wren's endearing tiny size
and lively character earned it a place on our smallest
coin, the farthing (one-quarter of an old penny).

A Wren is so small, it weighs less than a £1 coin. Little Wren is often described as mouse-like (as in the Brothers Grimm tale, page 161) – it hunts for food on the ground, or low down, close to the earth, hopping and flitting in thickets protected by bracken or bramble. In France they say the Wren is 'no bigger than … a mouse fart!'

LITTLE BIRD, BIG SONG

For such a tiny bird, the Wren has a huge voice. The Wren has the loudest song, in proportion to its body size, of any bird in Britain. It sings at 10 times the power of a crowing cockerel. The little bird vibrates with energy, its whole body aquiver, as its powerful song rings out, hence the collective noun: a chime of Wrens.

Like the Robin and the Tawny Owl, the Wren holds territory throughout the colder months, so listen for a Wren song brightening a winter's day.

Offer a Wren extra food by sprinkling grated cheese over the leaf litter, somewhere where hedges offer protection from predators.

THE BEST NEST

In spring, the male Wren builds a number of nests, each one an oval ball of old leaves, dry grass and moss, with a little round hole for a door. Wren nests are often low down (below knee height), hidden under hedges, behind ivy or in brambles. To court a female, the male builds up to 10 nests for the female to choose from. He takes the female on a tour of his creations, enticing her to view each one with a song, and with much hopping and wing-quivering. The female chooses the nest she thinks best, and tends to the soft furnishings, lining it with feathers.

The female broods the eggs herself, and, once the chicks are hatched, their father keeps a protective eye on his young and finds them food. Wrens often have two broods – the male may take the first brood to roost in one of the spare nests whilst the female incubates the second clutch.

Sometimes (where food is especially abundant), a male Wren might court two females, each with their own nest – he helps each partner defend her territory and helps feed both broods.

Wren families are famously large, hence the old nursery rhyme:

The dove says coo-coo, what shall I do?
I can scarce bring up two.
Pooh, pooh, says the Wren, I have got ten,
And keep them all like gentlemen!

A LIVELY LITTLE ONE

A Wren is a little ball of energy, always active and alert. Most often, you'll see a Wren only for a moment, flitting from one thicket to another. In flight, a Wren's short, rounded wings flap so fast that all we see is a flurry of feathers.

As winter days darken, garden birds spend more and more of the short daylight hours feeding, to stay plump enough to keep warm through the long cold nights. To survive the winter, the tiny Wren needs to feed almost all day long.

WINTER WARMTH

Freezing weather can be life threatening for small birds. A Wren's tiny size means it chills quickly. In cold winters, up to 80 per cent of the Wren population can die, as in the winter of 1962–1963, when snow fell for weeks, and drifts were deeper than a person is tall. Remarkably, with milder winters the followings years, Wren populations bounced back – thanks to their large families, Wrens remain Britain's most numerous bird.

How do tiny Wrens keep warm to survive the winter? They usually roost alone, but in freezing weather Wrens may roost communally, packed tight inside a sheltered space. An old nest offers a warm welcome in the cold, and Wrens roost together in both House Martins' mud-nests and squirrels' leafy dreys. One roost in Norfolk held over 60 Wrens at once. An old belief from Normandy was based on this observation: it was said that the cottage that gives shelter to a Wren's nest in the spring sees a miracle on Twelfth Night (5th January) – the male and female Wren return to the nest with all their children, to take part in the holy festival. If you have a nest box in your garden, after dark on a cold winter evening, peep (carefully) inside. You may see a feathery ball of Wrens cuddled up together, with heads facing inwards and tails pointing out. During the night, the birds change positions – taking turns in the cosy centre.

HOME IN A HOOD

A sweet, simple folktale from Brittany tells that a Wren nested in the hood of St Malo's cloak. The kindly saint didn't move his cloak until all the baby birds had fledged. Like Robins, Wrens do nest in all kinds of unusual places (such as a scarecrow's pocket, the base of a Magpie's nest, and trousers and jackets hung out to dry), so the story could well be true.

ON THE FEAST OF STEPHEN

Many folk-stories relate to the Wren's loud alarm call, and some offer a reason why the bird, by tradition, is hunted during the midwinter Wren Hunt ceremony (page 185).

Perhaps the best known is the story of St Stephen. Because of his Christian teachings, Stephen was imprisoned. He tried to escape whilst the guard was asleep but the jailor was woken by a Wren.

British ornithologist Edward A. Armstrong suggested that this folktale, expressing a tension between the Wren and Christian saints, suggests the bird was once revered by the Celts, and that its magical status was displaced with the coming of Christianity.

Another legend tells that Irish soldiers tried to sneak up on a Viking camp, where leftover breadcrumbs littered the top of a drum. When Wren pecked up the crumbs, the *rat-a-tat* woke the Vikings, preventing the attack. An almost identical folktale tells that in the same way, Oliver Cromwell, an English military leader who invaded Ireland, was warned of an Irish attack by a Wren.

In all these tales, the loud drumming that warns of attack is clearly related to the Wren's loud alarm call – its rush of rattling, trilling notes has the volume, pace and intensity of a drum roll.

KING OF THE BIRDS

The tiny wren is known as King of the Birds in countries all over Europe. The folktale that tells how Wren became king is well known throughout Europe. It has even been recorded across the Atlantic in North America.

The tale has truly ancient roots. Greek philosopher Aristotle, in his *History of Animals*, says that the Wren goes by the name of king, as does the Roman writer Pliny. The Greek philosopher Plutarch refers to 'Aesop's wren who was carried on the eagle's shoulders, then … flew out and got ahead of him.' Though the tale is not recorded in any surviving Aesop's tale, it was clearly known in ancient times. For at least 2,000 years, perhaps longer, the Wren has been King of the Birds.

There's a theory that the tiny bird the old tales originally referred to was actually a Goldcrest, a bird even smaller than the Wren (in fact, the smallest bird in Europe). Its bright golden crown (or 'crest') of feathers gives it its scientific name, *Troglodytes troglodytes*, and suggests that, originally, it was this tiny kinglet, rather than the Wren, who was king of the birds. The fact that the Goldcrest and the Wren were both known as 'kinglet' (in French, for example) might explain how the hero of the story changed over time, and the crown was passed from one bird to the other.

The Garryduff Gold Bird, a sixth-century brooch shaped as a Wren and decorated with scrolling wirework, was excavated in County Cork. Many believe the brooch is evidence of the importance of the Wren in ancient Ireland.

SACRED BIRD

The Irish word for Wren, *dreoilín*, is connected with the words for druid (*druí*) and magic (*draoi*, *draoicht*); the name is thought to be a contraction of the words *draoi* and *éan* (bird) – meaning druid bird. *Cormac's Glossary* states that the Old Irish word for Wren is *dreann* – 'druid bird'. A medieval Irish text, *Dreanacht*, 'Wren Lore', gives the prophetic meanings of Wren calls that come from different directions. A call from the north was a sign that someone dear to the hearer was coming to visit, a call from the east signalled the arrival of poets, but a call from the west foretold the arrival of wicked kinsmen.

In Irish accounts of the lives of the saints, the Wren is referred to as *magus avium*. A 'magus' is a magician or high priest (like the magi – three wise men – in the nativity story).

In Normandy, the Wren is named *la poulette au bon Dieu* – God's Little Hen – a title given to the bird because she was present at the birth of Jesus. She had her nest in the stable and brought moss and feathers to make a blanket for the holy child.

A Wren nesting near your home was thought to be good luck, and to kill a Wren is sure to bring misfortune:

Kill a robin or a wren,
Never prosper, boy or man.

A French folktale tells that a Wren brought fire from heaven for humankind. Another belief about the Wren's connection with fire from heaven comes from Pays de Caux, in France: if anyone robbed a Wren's nest, their own house would be struck by lightning.

'ALTHOUGH HE IS LITTLE HIS HONOUR WAS GREAT' – THE WREN HUNT

The Wren Hunt is perhaps the longest-surviving bird ritual in Europe, a ceremony with ancient roots. Edward A. Armstrong even put forward a theory that the Wren cult reached the British Isles during the Bronze Age, carried by megalithic builders along prehistoric trade routes. The custom was common across England and Ireland (except in Ulster), as well as in Wales, France and the Isle of Man.

The midwinter ritual took place within the twelve days of Christmas, usually on St Stephen's Day (26th December), Twelfth Night (5th January) or Epiphany (6th January). It usually involved the males of the community ('Wren Boys') hunting a Wren and decking it in evergreens, and parading with it from cottage to cottage, dressed up in rag-tag costumes and singing songs. Sometimes, Wren Boys were given food, drink or coins:

The wren, the wren, the king of the birds,
St Stephen's Day was caught in the furze,
Although he is little, his family is great,
I pray thee, good landlady, give us a treat.

IRISH WREN HUNT SONG

In Llanidloes in Wales, a live bird was carried in a 'Wren house' made of wood with doors and windows, or a stable lantern decked with ribbons. In Pembrokeshire at Epiphany, a boy came round with coloured paper streamers in his hat and the Wren in a cage, singing:

Come and make your offering,
To the smallest, to the king.

In Pembrokeshire, after the procession, the bird was set free. On the Isle of Man, the Wren was killed, and, after the procession, buried with solemn ceremony in the churchyard. Another medieval custom of the twelve days of Christmas was the appointment of a Lord of Misrule – a young boy was given the role of king, in a reversal of the everyday. The ceremonial Wren Hunt during the twelve days of Christmas is a similar reversal – at all other times, it is unlucky to kill a Wren. Armstrong proposes that this, as well as the fact the bird was given a royal title, suggests that the Wren was once revered. Though the origins of the tradition are unknown (and debated), many writers have expressed the idea that the ritual death of the Wren may even be a relic of a pagan rite of sacrifice and rebirth. It is easy to picture the earth-dwelling Wren as a symbol of the downward-moving energy of the dark half of the year, which dies at midwinter when the light is reborn.

In 1848, to prevent cruelty to animals, the mayor of Cork forbade the killing of a Wren during the festival. One hundred years later, in the 1950s, the Wren Hunt was still being celebrated in the south of Ireland, using an effigy to represent the Wren (a potato stuck with feathers, for example). 'Wran's Day' is still celebrated in Ireland today, with great exuberance. And the tradition has been rekindled on the Isle of Man, where a stuffed Wren is used each year, and, instead of feathers, ribbons are taken from the 'Wren Pole' for luck.

On the Isle of Man, the reason for the Wren Hunt is explained in the story of a sea spirit, always attended by storm and shipwreck. Fishermen tried to catch her, but she escaped by turning into a Wren. Since then, she's had to take the form of a Wren every New Year's Day. As a result, Wren feathers were believed to offer protection from storms, and were treasured as talismans against shipwreck.

In many areas, great play was made of the bird's tiny size, with men pretending to labour under the heavy weight of the Wren they carried. A cart is needed to carry the Wren in the rollicking song sung on the Isle of Man – a culminative (35 verse!) call-and-response:

> *We'll away to the wood, says Robin to Bobbin,*
> *We'll away to the wood, says Richard to Robin,*
> *We'll away to the wood, says Jack of the Land,*
> *We'll away to the wood, says every one.*
> *What shall we do there? says Robin*
> *to Bobbin … etc*
> *We'll hunt the wren, says Robin to Bobbin … etc*

OPENING OF *THE HUNTING OF THE WREN*, COLLECTED BY WILLIAM HARRISON, FOR THE MANX SOCIETY, 1843

CELEBRATE THE WREN!

Create a midwinter tradition for your family, in celebration of the Wren, on St Stephen's Day, Twelfth Night or Epiphany. By tradition, Epiphany is the night the three wise men arrived at the stable with their gifts for Jesus. In European countries (and, before the Industrial Revolution, in Britain too) the festival was celebrated with a special Three Kings Cake. Inside the cake is hidden a bean (or coin) – whoever finds it in their slice is crowned king. (You can buy a Farthing online for less than a pound, if you want to make a Kings Cake with a 'wren coin' inside.) Whoever is crowned could choose an activity for the whole family to enjoy together.

You could also use a farthing (or a home-made pom-pom Wren) to play 'Hunt the Wren' in the same way as Hunt the Thimble (one person hides a small object, and the others try to find it, to cries of 'cold', 'warmer', 'HOT!').

'KING OF THE BIRDS' A FOLKTALE FROM WALES

A story that celebrates the Wren and highlights it's short, turned-up tail

Once upon a time all the birds of the sky gathered together to choose a king. It was agreed: whoever could rise the highest in the sky would be crowned.

They were all there, little Blue Tit and Robin Redbreast, Goldfinch and Green Woodpecker. Even Tawny Owl, who usually slept all day, was there, yawning and blinking her sleepy eyes. And the smallest of them all was Wren, a little ball of feathers, no bigger than a mouse.

The biggest was Eagle. 'None can challenge me,' Eagle declared.

'I challenge you!' piped up a little voice. Wren was bobbing up and down with excitement.

With piercing eyes and frowning brows, Eagle stared down at Wren. 'You? Impertinent bird! Why, I've never seen you hop higher than a hedge.' Pinned in Eagle's stare, Wren couldn't help but quiver. But he spoke up, loud and clear, 'I hop very well. And I'm good at hiding!' Eagle sniffed, and turned away.

Just in time. Ready ... Steady ... GO!

There was a great flapping and fluttering as all the birds rose into the air.

Up flew Lark, straight up, singing himself higher and higher.

But Raven flew higher still, delighting in the heights, doing daring dives and twisting tumbles. He even rolled right over, flying upside down!

The little birds, Blue Tit, Robin, Goldfinch and all the rest, were soon left far behind. One by one, they dropped back down to earth.

But highest of them all soared Eagle. High, high, high into the sky, riding the wind. On long, strong wings he soared, high as the sun.

Eagle looked down with his sharp eyes. 'Well, little Wren, where are you now?'

'Here!' came a cry, and up sprang Wren, up from Eagle's feathers, where he had been hiding all along. He was so little and so light, Eagle had never even known he was there. Wren was bouncing with energy. He hopped upwards, a blur of whirring wings, a good hedge-height higher than Eagle.

'I'm the King of the Birds! I'm the King of the Birds!' piped Wren.

Eagle couldn't fly an inch higher. He was exhausted. And he was *so* cross. When Wren flew past, on his way back down, Eagle swung at the little bird with his wing. Wren went tumbling to the ground.

... And landed – *BUMP*!

He bounced right back up again, full of energy. He wasn't hurt, but his tail feathers were bent and broken – the tips were snapped clean off.

And that is why, from that day to this day, the Wren has a short little tail. But he sticks it straight up, proud as can be. After all, he is King of the Birds.

There are many variants of this tale. In Irish and Norwegian versions, Eagle is so cross with Wren he throws him to the ground, and when he lands – bump – the tip of his tail breaks clean off. Hence Wren's short little tale.

Another ending says that Eagle struck Wren with his wing, and so, even today, the injured Wren can only flit from hedge to hedge, never rising higher than a hawthorn bush.

In a Welsh version, the birds collect their tears to drown Wren. I like the way Rosalind Kerven re-imagines this; in her version for children, the birds are upset because their new king is injured. They gather herbs to add to the pot of tears, to make a brew to heal Wren's tail. Owl knocks over the pot, and the brew is spilled – Wren's tail has been short ever since.

MAKE A WREN FOR ST STEPHEN'S DAY OR TWELFTH NIGHT

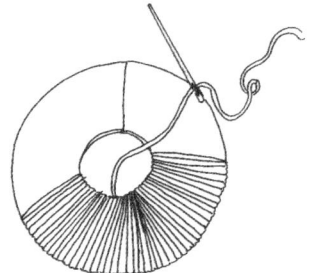

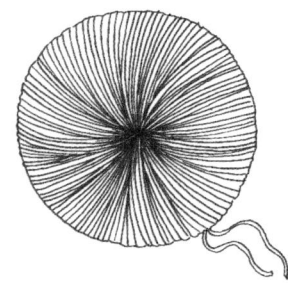

Making a woollen pom-pom is a warm and restful activity for little ones on a winter afternoon. A pom-pom is easy for young children to make, and is simple to transform into a little round wren.

Make a Wren on St Stephen's Day (26th December), to hang upon the Christmas tree, in honour of the King of the Birds.

Pom-pom Wren

You will need:

- Yarn in a Wren-like shade of brown (wool is best, as it biodegrades, unlike synthetic yarn)

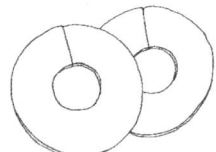

- Two cardboard rings
- Scissors
- A tail shape cut from a piece of woollen felt
- Two small wooden beads
- Two small triangles of woollen felt
- Needle

1 Cut out two rings of cardboard.
2 Put the rings together and wind the yarn around them firmly and evenly (children could use a blunt tapestry needle). Keep going until the central hole is completely filled.

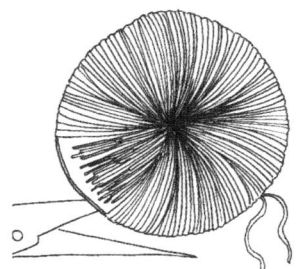

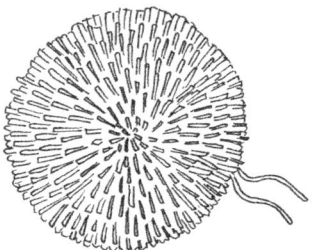

3 Using sharp scissors, cut the yarn along the circumference of the ring, with the scissor blade between the two pieces of card.

4 Take a length of yarn and thread it between the two pieces of card. Wind it around the centre a few times, pull tightly and tie into a knot. Attach a length of yarn for hanging.
5 Cut away the card.
6 Fluff up the pom-pom. If you like, trim the pom-pom into a neat ball shape.
7 Glue on the two triangles of felt to form the beak.
8 Glue on two beads for eyes.
9 Position your felt at the other end of the pom-pom to make a tail. Glue the tail on so that it sticks straight up.

FURTHER RESOURCES

House Martin Conservation UK & Ireland
housemartinconservation.com

Swift Conservation swift-conservation.org

Wetlands and Nature Reserves
To find the nature reserve or wetland nearest
to where you live, try:

- Wildfowl & Wetlands Trust (wwt.org.uk).
- The Wildlife Trusts (wildlifetrusts.org).
- The RSPB (rspb.org.uk).

Starlings in the UK
For more information on current locations
of Starling roosts in the British Isles, see:
starlingsintheuk.co.uk

Moors for the Future
For a printable moss ID guide,
see the Sphagnum ID Guide at:
moorsforthefuture.org.uk

USEFUL ORGANISATIONS

RSPB rspb.org.uk
British Trust for Ornithology (BTO)
bto.org

Reference List
Sources notes and full references can be found
on the author's website: dawncasey.co.uk

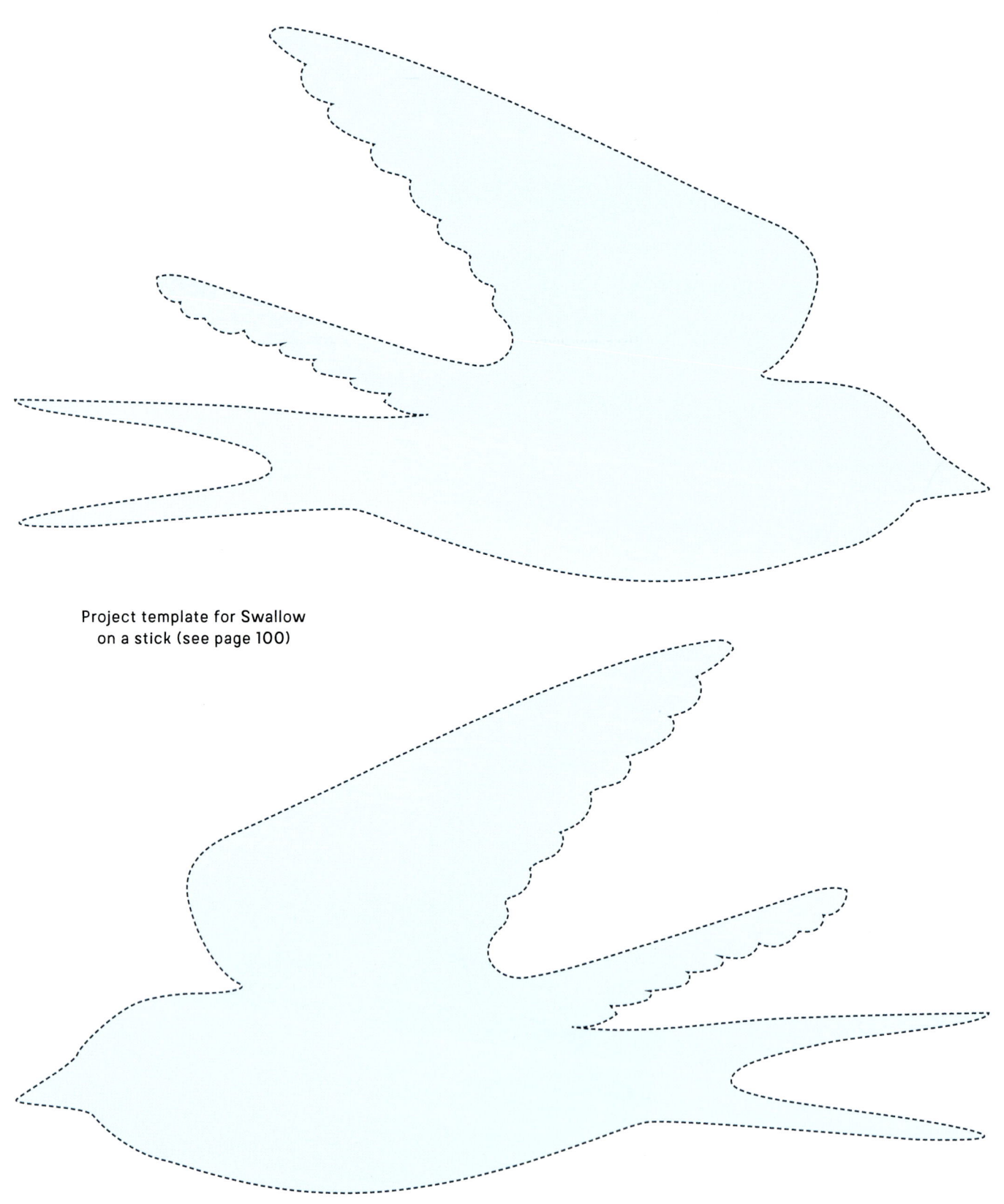

Project template for Swallow
on a stick (see page 100)

INDEX